PILLBUGS AND OTHER ISOPODS

Cultivating Vivarium Clean-Up Crews and Feeders for Dart Frogs, Arachnids, and Insects

Orin McMonigle

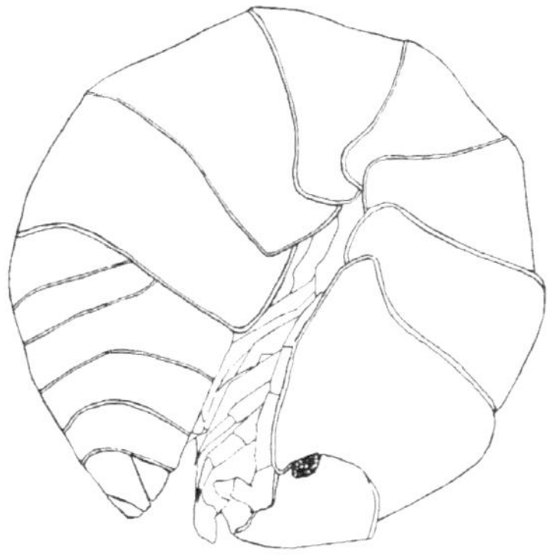

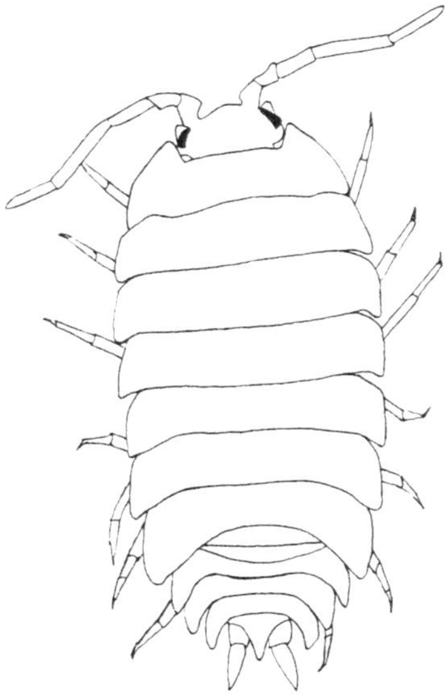

COACHWHIP PUBLICATIONS
GREENVILLE, OHIO

ACKNOWLEDGEMENTS

I would like to thank Ferenc Vilisics, Christian Schwarz, and Tammy Wolfe for editorial suggestions. I would also like to thank Christian Elowsky, Ingo Fritzsche, Peter Clausen, Alex Yelich, and Ron Wagler for input. I thank my wife and daughters for their support and God for creating such varied fauna and my interest.

Pillbugs and Other Isopods

ISBN 1-61646-207-8
ISBN-13 978-1-61646-207-9

CoachwhipBooks.com

Front cover: *Armadillidium maculatum* and *Armadillidium nasatum* "peach" form.
Back cover (Clockwise, from Top Left): Mating pair of *Armadillidium maculatum, Porcellio scaber*, group of calico *Porcellio, Porcellio dilatatus* getting ready to molt.

CONTENTS

Armadillidium maculatum and *Armadillidium nasatum* "peach" form.

INTRODUCTION

Isopods include fascinating creatures that capture the hearts and minds of young children. The "roly poly" may be the favorite creature of humans below the age of six. Literature strongly reflects this inclination since there are quite a few children's books on keeping the roly poly as a pet, but none for adults. One of the earliest, *The Pillbug Project: A Guide to Investigation* (Burnett 1992) presented isopods as an ideal investigative project for children. More than a dozen children's books have been written about them,

Kids are fascinated by isopods.

mostly in the last ten years. Subsequent books have likewise been designed to introduce youngsters to beginning biology but there is also recent children's fiction employing them as characters, such as *Roly Polies* (Carretero 2011). Although pillbugs are the ultimate real transformer, only one *Transformer* from the 1990s *Beast Wars* line was an isopod (the toy rolls up into a sphere the size of a baseball). A great variety of small cages are manufactured and sold for children to place their finds in when collecting bugs. These most often house isopods since they are very common and easy to capture. At least one of these popular toys is a dedicated isopod enclosure, the *Roly-Poly*

Roly Poly Playground™

Playground (2012). A pair of acrobatic roly polies made appearances in *A Bug's Life* (1998) as they transformed from ball to bug in routines. *Starship Troopers 3: Marauder* (2008) hosted giant pillbugs the size of basketballs that exploded when they uncurled, called "Kamikaze bombardiers." This movie was not meant for children, but certainly was meant to appeal to the boy inside the man and the original book by Heinlein (1959) was not written for adults.

There are various scientific papers on the biology, morphology, and systematics of the terrestrial isopods, but nothing for the adult enthusiast. Other than fascinating biological data mostly slanted towards marine species, there is almost nothing between childish naiveté and scientific drudgery (fascinating scientific discovery being something else entirely). Compared to other invertebrates regularly kept in captivity, isopod literature is amazingly sparse. Isopods are commonly left entirely out of entomology books because they are not insects, and they also rarely show up in arachnid books or combinations of the two. Some identification and list books on terrestrial invertebrates present one to three species and may (rarely) devote a full paragraph to the entire order. Teen- to adult-level book references to husbandry have included the two short paragraphs in *Rearing Insect Livestock* (Dunn 1993) and what may be the most thorough treatment of captive isopods at barely three-and-a-half pages in *Invertebrates for Exhibition* (2011). Various feeder texts mention the great value of isopods in feeding certain animals. The earliest reference to specific uses for feeding invertebrates is probably *Assassins, Waterscorpions, and Other True Bugs Care and Culture* (2001), while the most extensive feeder references are a few pages on use with tropical fish in *Encyclopedia of Live Foods* (1975) and for herps in *Rearing Live Foods* (2009). There are a total of three moderately in-depth articles from *Invertebrates-Magazine* but with exception of *Bugs Das Wirbellosen-magazin* (1:1, 2013), a new, German invertebrate magazine containing a picture and paragraph for the spectacular *Porcellio* cf. *bolivari* as part of a larger article, and an old American Tarantula Society *Forum Magazine* (10(1), 2001) reference to their confusion with giant pill millipedes, I have never seen them in entomological periodicals. Even if all the above literature was combined into a single resource it likely would not take half an hour to read through (and if you threw out the *Invertebrates-Magazine* articles there might be ten minutes left). This book is intended to remedy the lack of printed material available to the adult isopod hobbyist. Details will be concentrated on those species kept in culture with the known history of captive stocks and specific uses as terrarium clean-up crews. Hopefully this text can combine the excitement of a three-year-old with scientific data and not smother or bury either.

Retrax, an evil transforming robot pillbug.

Why isopods? They are neat little creatures. They require very little care, do not need special temperatures, and are easily and cheaply housed. Most people find captive

Not only do isopods make a great terrarium cleaning tool
but also a great food for other crustaceans.

One of the amazing blue iridovirus *Porcellio* sp., photo taken in 2004. © Alex Yelich

Even this very large *Porcellio dilatatus* is small by pet bug standards.

species take little effort to culture and required feeding frequency is monthly or less. They are completely harmless to humans and will dry up quite rapidly if they get loose in the house. They make excellent clean-up-crews for small terraria and can be incredibly useful feeders for certain amphibians, predatory bugs, and omnivorous crustaceans. Lastly, most species have uncommon, though not terribly rare, color forms that can be adapted through unnatural selection like fancy guppies or miniature freshwater shrimp.

Three of the more fantasticly colored cultivars.

Captive culture of isopods is relatively new. Some of the biological supply houses have offered terrestrial and aquatic isopods for decades, but these were primarily wild-caught and the intent short-term classroom study. Only two of the commonly traded culture stocks, (Spanish) orange isopods and white "micropods," have roots in the ultimate years of the prior millennium and even these were rarely maintained more than ten years ago. The oldest continuously maintained stock with no introductions of unrelated animals I know of is a culture of *Armadillidium vulgare* from 1995. It would take very little effort to keep a stock for many decades, so lack of any really old stocks refers only to the previous lack of interest in isopods. In the last few years, additional species useful for terraria and a handful of captive cultivars have arrived on the scene. Increased interest can be mostly attributed to wide use in dart frog habitats, but there are a number of other applications and some of the prettier forms may be kept for their own sake.

Usually when I put together a text on rearing a type of invertebrate it is not because anyone has requested it but because it is the sort of text I would be interested in reading but cannot find. I have thought about writing on isopods ever since I acquired fantastically bright blue ones in the late 1990s, but had been waiting for one large, very easily reared, and showy species to build the text around. That species has not yet materialized. However, slowly over the years a number of very interesting species and cultivars have come to be commonly reared. Also, this is the only time I have heard from one or two people asking about availability of a care guide before it was written. (I am never 100% sure a single person will be interested in a copy.) I had finally decided it was time to assemble everything I knew about captive isopods two years ago, but was sidetracked by larger Coachwhip Publications projects. This text brings together culture methods, biological details, and information on species and cultivars presently being maintained by enthusiasts. I worry that some new fantastic species will enter captive culture and make this text out of date—however, it is also something I greatly hope for. This expanded hardcover edition not only includes more information and a legion of photographs that simply could not be crammed into the small guide, but also includes pictures and details for the spectacular *Porcellio* cf. *bolivari*, *Ligia exotica*, *Venezillo parvus*, and recently arrived *Armadillidium maculatum*.

Bathynomus species. (cc) Expedition to the Deep Slope 2006, NOAA-OE

1
RELATIVES, MORPHOLOGY, AND NAMES

THE ORDER ISOPODA is a large, diverse group of arthropods from the subphylum Crustacea whose members are more often thought of as seafood than as "bugs." The Crustacea includes a number of shrimp-like creatures from tadpole shrimp and brine shrimp to mantis shrimp and true shrimp. The class Malacostraca includes the most familiar crustaceans such as crabs, lobsters, shrimp, and isopods. Freshwater crayfish, various shrimp, and a few crabs have become increasingly popular aquarium pets in recent years. Interest in crayfish and miniature shrimp somewhat parallels the isopods since much of the interest has come from rearing and developing attractive color forms under captive conditions. Orange, blue, or white crayfish forms are seen across various genera and species, while the colors and patterns seen in multiple varieties of a few common, miniature shrimp are far more developed.

The malacostracan superorder Pericarida includes the Isopoda, Amphipoda, Mysida, and a few lesser-known crustacean orders. Amphipods are the common scuds from marine aquaria, but they include freshwater scuds, tiny hopping terrestrials known as grass shrimp, and beach hoppers. Mysid shrimp had been a popular prey item cultured for seahorses and other picky feeders maintained in marine aquariums. Mysida are also known as opossum shrimps because, like all the Pericaridia, they hold eggs and young in a brood pouch or marsupium and the immatures are somewhat large and look like the adults (versus the nearly microscopic planktonic larvae of most true shrimp that bear no resemblance to the adult animal).

Within the order Isopoda there is incredible diversity. There are presently more than 10,500 species known (McCormick 2007). These are distributed in eleven different suborders. The largest suborder, Oniscidea, is what we normally see, but the long-legged aquatic *Asellota* can also be encountered in many areas. Isopods are the only arthropod order to conquer virtually every habitat type on earth.

A little over half, more than 5,000 species, are found throughout the ocean from the deepest trenches to tide pools. The largest known are among the most common of all deep sea denizens, *Bathynomus* spp. from the suborder Cymothoida. The largest of all known isopods, *Bathynomus giganteus*, can reach an impressive 7.5″ to 14.5″ (19 to 37 cm) though some unverified, whimsical fish stories suggest *Bathynomus* species can exceed two feet. Specimens trapped in deep waters off the coast of Australia, India, and Mexico have proven virtually identical (Caldwell 2008). It may have been the first isopod ever to make its way to television. *The Living Planet* (1984) incorrectly calls *Bathynomus* an amphipod and claims it is related to horseshoe crabs, which are not

This blue *Procambarus alleni* female and her young are permanently blue due to a nonlethal iridovirus. Crayfish and other decapods carry the eggs and young on the pleopods rather than below the pereon and there is no protective marsupium.

This small *Neocardia* sp. shrimp is a form selected for red coloration known as the cherry shimp. This specific form started the mini-shrimp craze.

Crustaceans such as this male fairy shrimp come in a wide variety of forms.

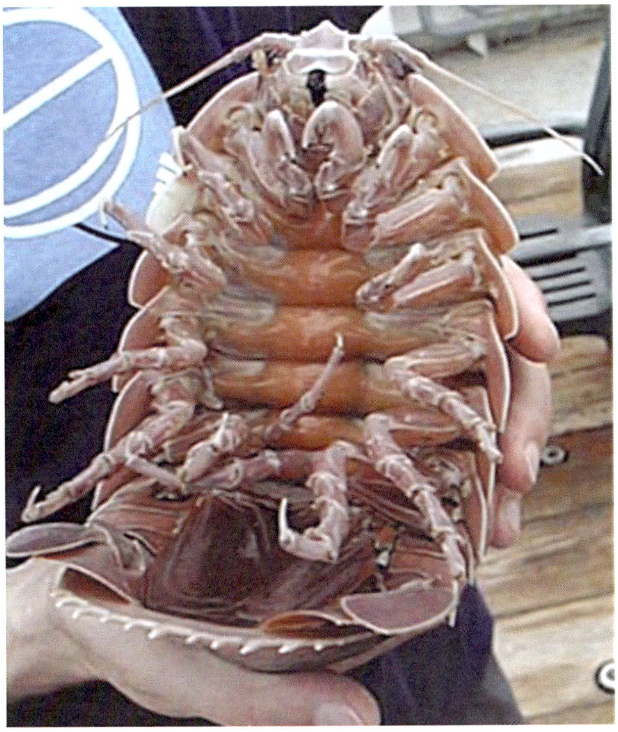

Bathynomus giganteus would be a little too large to make a good cage pet even if it could be readily cultured. NOAA

crustaceans. More recently, deep-sea isopods were fictionally used by NASA to fake alien fossils in the pop-fiction *Deception Point* (Brown 2001). A number of public aquariums have tried to keep *Bathynomus* since the late 1980s (a presentation by Shedd or Baltimore Aquarium to Cleveland Saltwater Enthusiast's Association circa 1988 included a preserved specimen) but to this day none have seen anything beyond short-term survival (pers. comm. Caldwell et al.).

Various marine isopods, primarily those also in the suborder Cymothoida, colonize the bodies of fish, shrimp, and related animals as parasites for at least part of their lives. The most unusual and intriguing of all isopods is one that lives in the mouths of fish and is found in waters from the coast of California to South America. This species, *Cymothoa exigua*, parasitizes the tongue, where it lives from the fish's blood. The tongue eventually dies off due to physiological stress, which normally would doom the fish. However, *C. exigua* then takes the place of the tongue and becomes a fully functional replacement (Walls 1984). This is the only case known in which a parasite fully replaces an organ of its host. One look in the affected fish's mouth reveals a crazy looking tongue with legs and a pair of eyes looking back at the observer. This inspired a 2012 horror flick featuring parasitic isopods as villains but the most interesting aspect is unfortunately lost to cheap gore.

The largest terrestrials are far too small to make a good replacement tongue.

There are some marine isopods growing as large as 12″ (30 cm) that can be found in shallow waters, but these large species have not been seen in the marine aquarium hobby because they are from the frigid waters of

Ocellaris clownfish, *Amphiprion ocellaris*, with tongue biter isopod. © Jeremy Keith Brown

Fish louse *Anilocra* sp. on narrowstripe cardinalfish, Lembeh Strait, Indonesia. © iFish

Antarctica. A whole host of other crustaceans such as spiny lobsters, slipper lobsters, mantis shrimp, hermit crabs, etc. are popular fares in marine aquaria but these large handsome specimens can be kept near room temperature and near our normal atmospheric pressure. A variety of very small tropical species, from those with short legs and fat bodies that possess lobster-like swimming tails to those resembling house centipedes with extremely long legs and thin bodies, have long been common imports on "live rock." These can be important, or at least significant, inhabitants of mini-reef

House centipedes superficially resemble some of the long-legged aquatic isopods but have only one body section and conspicuous maxillipeds.

aquaria. (For years I would see a small asellotid in every pet shop and my own reefs, though I have not seen that species in a while.) I had a fascinating member of the suborder Sphaeromatidea in one of my reef tanks for three or four years. It superficially resembled the familiar terrestrial species but the males had massive, spiked uropods and though it walked like common terrestrials, it would swim in rapid bursts when disturbed. A variety of species are very common in sea grass beds and among marine macroalgae (seaweed). In June 2010, off a beach in Norfolk, Virginia, I collected a tiny ~30mm piece of sea lettuce to see if it would grow in my reef tank. Wave action tore this tiny fragment and many others off large plants attached to rocks somewhere offshore. I grabbed the tiny, flat blade between a finger and thumb as it floated offshore. (Washed up pieces were ignored since sea lettuce is dead within minutes of washing up even if it appears live.) One of the smallest pieces was chosen since it would be much easier to keep alive for a few days in transport. The next day I was shocked to discover a rather large (28 mm x 7 mm) bright green isopod from the suborder Idoteidae in the small deli-cup. This and similar looking species on the west coast are probably incredibly common but rarely noticed even when they wash up because they perfectly match the color of the seaweed they live on. Isopods are not just washed up on shore, they are often the dominant fauna of many shoreline habitats—some look very similar to inland terrestrial species.

There are nearly 4,000 terrestrial species (Schmalfuss 2003) though there may be some debate on aquatic vs. terrestrial status for certain species found where land and sea meet. Further inland, the isopods are overshadowed by various larger life forms but are still the most common fauna in my yard and the nearby forest (with exception of microscopic or near microscopic mites, springtails, protozoans, and bacteria). Although suited to moist habitats, they can be common in New World and Old World deserts. *Venezillo arizonicus* is native to the Sonoran and Colorado deserts, the driest and hottest areas of the United States. In the Sonoran Desert, isopods are encountered in far greater numbers than desert cockroach nymphs (Polyphagidae, *Arenivaga* spp.) among mesquite leaf litter accumulations. One large species encountered in Arizona appears

Idotea stenops. (CC) Jerry Kirkhart

Idotea (Pentidotea) wosnesenskii. (CC) Jerry Kirkhart

Ligia pallasii, Alaska. (CC-SA) D. Sikes

Mate guarding in *Armadillidium maculatum*.

to be *Porcellio laevis*, though it borrows and is highly resistant to desiccation (McMonigle 2004). *Hemilepistus* species look similar to the well-known *Porcellio* (and were considered members of the genus *Porcellio* for nearly a hundred years) but have conquered Old World deserts mostly through behavioral modifications. High population density means they are often the most successful inhabitants of this environment (Linsenmair 1974). A number of common inland species are found in oceanside sand dunes that share some characteristics with desert areas. Some species like *Armadillidium album* are restricted to this habitat (Nichols et al. 1971). Terrestrial isopods are found in habitats with moisture gradients between swamp and desert, but also in caves and the frozen moisture of the subarctic.

Though far less common than marine and terrestrial species, the 500 or so freshwater species are mostly disproportionately long-legged asellotids that barely resemble terrestrials. There are also a number of peculiar eyeless freshwater species in Australia and New Zealand from the suborder Phreatoicidea. Although they do not look so incredibly different, few people would recognize any of the freshwater aquatics as roly poly relatives.

Armadillidium nasatum with a nose-like horn (purplish form brownish cull).

BODY PARTS

The isopod body is divided into the usual three regions (head, thorax, and abdomen), but the main body divisions do not match perfectly so they are named differently (head, pereon, and pleon). The head of the isopod is not attached to a large carapace as with many crustaceans or most arachnids. As with centipedes and most crustaceans, there is an extra pair of mouthparts called maxillipeds because they are adapted from the front pair of legs. However, unlike lobsters or centipedes where maxillipeds come from a segment behind the head and look like adapted legs, isopod maxillipeds do not betray their origin. In most isopods this means the head is imperceptibly fused to the first (not the first visible) thoracic segment. Since the first segment is fused to the head, the thorax does not perfectly match the visible body divisions, so the next seven segments of the thorax are together termed the pereon. Each of these large segments has a pair of legs called pereopods. All the terrestrial species and most aquatics have fourteen legs, not including maxillipeds. The large pereon (thoracic area) looks different from primitive insects because the upper surface in most species is divided into seven equally large plates rather than three. Also, the thoracic plates seem to flow into the plates of the abdomen at a glance, so they are often confused with the millipedes that do not have a divided thorax and abdomen.

The abdomen of crustaceans is also known as the pleon because this division also includes a thoracic segment. The crustacean pleon is often referred to as the tail (in the case of lobster and shrimp tails specifically). The upper segments are called the pleurites while the terminal, tail-like, and commonly triangular segment is known as the telson. A pair of small dorsal appendages called uropods (or stylets) are the segments on either side of the telson,

There are long spine-like uropods and five lung pairs down both sides of the pleopods for *Cylisticus convexus*. (The lungs of *Trachelipus* and *Venezillo* appear similar.)

Lungs, two pair, on pleopods of *Porcellio dilatatus*, all *Porcellio*, *Armadillidium* species, and others.

Ommatidia that make up the eyes of some isopods can be so small and numerous they cannot be seen without a microscope, *Ligia exotica*.

but these are expanded into large, widened plates in crayfish, shrimp, swimming isopods, and many swimming crustaceans. In most terrestrial species the uropods are long and thin and may have a sensory function somewhat like the cerci of insects. The rows of flattened appendages on the underside of the pleon are known as pleopods (also called swimmerets). As with many other crustaceans, the male pleopods are secondary sexual characters, but unlike with familiar crabs and lobsters, the female's are not adapted for holding eggs.

The pleopods can be used in swimming but the primary use is for breathing. Other familiar crustaceans have gills on the thorax (lobster, crab, etc.) or on the legs (brine shrimp, triops) so the abdominal pleopods can be adapted for holding eggs or swimming. Aquatic isopods normally pulsate the pleopod gills to keep water flowing across. Above the water these gills can work if kept moist—littoral species wet them with water from time to time. Most of the familiar terrestrial species have "lungs" on the pleopods. These show up as white markings or squares that can be helpful in certain identifications. Whether they have lungs or not, terrestrial species are resistant to drowning and may walk around for hours submerged (but will eventually drown).

Crustacean eyes are far more variable than those of other arthropods. Isopods run the gamut. There are eyeless species, but in captivity they run from the single or triple ocelli found in the micropods, to eye patches of most large terrestrials, to the spectacularly complex compound structures seen on our slaters, including the huge *Ligia* species. Some crustaceans, such as

Isopod face SEM Jungle micropod, a single ocellus and reduced inner antenna pair are visible here.

Eye patch *Porcellio dilatatus*. © Henry Kohler

Clusters of around a few dozen ocelli make up the eye patches of many large terrestrials, *Cylisticus convexus*.

Porcellio spinicornis spines on antennae. © Henry Kohler

Porcellio dilatatus has a telson with a rounded tip, white form. © Ron Wagler

This pale section *Porcellio* sp. does not have a rounded telson like *P. dilatatus*, it is not smooth like *P. laevis*, not heavily tuberculated or as small as *P. scaber* nor is there a spine on the antenna like *P. spinicornis*.

members of the Stomatopoda (mantis shrimp), that also have impressive compound eyes, can have at least sixteen different types of color receptor cells (compare to humans with three and butterflies at five) (Chiao et al. 2008). Each stomatopod compound eye is divided with sections that detect polarized light and depth perception (Tait et al. 2005). These crustaceans are often considered to have the best eyesight in the animal kingdom. Nevertheless, while stomatopod crustaceans (and dragonflies) top out at around 3,000 ommatidia, *Bathynomus giganteus* eyes contain nearly 4,000 facets (Caldwell 2008).

Four antennae and biramous uropods are easy to see on most aquatics, *Asellus* sp.

Unlike insects with two, crustaceans normally have four antennae. This is true for isopods but the inner pair for most terrestrials is so reduced they are difficult to locate without a microscope. The last segment of the antenna is called the flagellum. It is usually composed of two or three articles for common terrestrials, but can be comprised of a dozen or more bead-like articles in shore-dwelling *Ligia* spp., (and also members of the families Trichoniscidae, Mesoniscidae, Styloniscidae, etc.). The flagellum often ends in terminal setae, which can help in identifying a species.

Another feature that sets crustaceans apart from insects is the presence of branched appendages. Whether an appendage is branched or unbranched may be seen labeled as *uniramous* (uni = "one", and ramous = "branch") and *biramous* (bi = "two"). Branching is not very noticeable in familiar isopods since the biramous pleopods are small and closed, but the branched uropods of sea slaters and *Asellus* are easily seen from above.

Immature *Ligia pallasii* are a handsome green and white mottled pattern. One is molting here and the colored rear half is apparent below the loose grayish shell.

TAXONOMY AND SYSTEMATICS

Isopod taxonomy can be easy or difficult to follow depending on knowledge of specific morphological characters and the number of species. In northern areas the lack of variety usually makes identification by a few key features possible, but if you do not know these features you could spend hours with a quality stereo microscope and discover nothing. Isopod taxonomists concentrate on certain groups or regions, so finding one who could identify a small Central American species such as the "jungle micropod" even to the generic level with certainty is beyond difficult.

In the case of our three most common volvating species, identification is very easy.

The St. Helena spiky yellow isopod, *Pseudolaureola atlantica*, is found only on that South Atlantic island. It is threatened by habitat loss due to invasive plants. © Mike Thorsen Imagery

Madison Cave freshwater isopods, *Antrolana lira*, devouring shrimp bait during a survey in Virginia. These are a Federally threatened species. USFWS

Armadillidium nasatum has a large, upturned plate between the eyes (in other animal species with a nose-like process the name is spelled *nasutum*), *A. vulgare* has no such process, and *Cylisticus convexus* has spiky uropods that look nothing like the flattened plates of the previous two. (*Armadillidium* tails resemble a lobster's.) There are a number of other differences but these alone make identification very easy. In Europe there are a number of additional *Armadillidium*, but most are similar in size and color to variation common within the immature color range of *A. vulgare*. Other, less common volvating species are found in the genus *Venezillo*. There are eight U.S. species, of which only the tiny *V. parvus* is introduced (Jass and Klausmeier 2001). They are easy to differentiate by locale because they are not common everywhere, but are most easily separated from other genera because the telson ends in a wide blade rather than a flat point. (In addition, there are five lung pairs like *Cylisticus* rather than the two pairs like *Armadillidium*.) There are also a few native *Armadillo*, *Armadilloniscus*, and *Tylos* species, but these are extremely rare. The *Armadillidium* are members of the family Armadillidiidae, while *Armadillo* and *Venezillo* species are in the family Armadillidae, *Armadilloniscus* from the family Scyphacidae, and *C. convexus* is from the family Cylisticidae—all from the infraorder Ligiamorpha, while *Tylos* are from an entirely different infraorder, Tylomorpha (Jass and Klausmeier 2001).

Uropods and blade-tipped teslon of a *Venezillo* sp. found in the Florida Keys.

When it comes to representatives of the family Porcellionidae, specifically *Porcellio, P. laevis* has a smooth carapace, while *P. spinicornis* has a lumpy carapace like *P. scaber*, but its color is mottled with yellow patches, while the head (bearing a prominent middle lobe) always remains dark in color. Also, there is a small spine projecting forward on the third antennal segment and the face is indented. The vast majority of "*P. scaber*" in my yard are actually *Trachelipus rathkii* (family Trachelipodidae), because though they have the same looking head, antennae, and overall body features, if you flip them over and look at the pleopods there are five small white patches rather than two larger ones along each side (these are the lungs). *Porcellio dilatatus* are easy to differentiate from all the above because they have a rounded rather than pointed telson. The four native U.S. species of *Porcellio* can't even be found in photographs. *Oniscus asellus* (family Oniscidae) again looks similar to *Porcellio* and remarkably similar to some color forms of *P. spinicornis*, but is easy to pick out because there are no white areas on the pleopods and their flagella has three articles, while *Porcellio* and *Trachelipus* species have two. *Porcellionides* species tend to be a little smaller and the abdomen is notably narrow compared to the body. Their body color is brown to gray without prominent markings. *Philoscia muscorum* (family Philosciidae) are difficult to confuse with the above since they have a smooth surface and unique spotting pattern in addition to a tapered abdomen, but they look similar to other members of the family Philosciidae, including *Atlantoscia* and *Floridoscia* from Florida and even the various *Littorophiloscia* from coastal areas. Philosciidae can also be confused with small sea slaters in the genus *Ligidium*, but as with the large *Ligia* species, sea slater eyes are more complex and the last antennal segment is broken up into a number of small articles, very unlike the three of *P. muscorum*.

Pillbugs commonly volvate but often uncurl rather quickly, *A. nasatum*.

When it comes to the "micropods," the white micros are the only ones traded with even a generic name, but that is because *Trichorhina tomentosa* is a common small white species, not because it was originally keyed out by a taxonomist. There are some tiny, temperate *Trichoniscus* species found mostly in the eastern states that may be confused only in that they also reach only a few millimeters at adulthood. However, they are mostly smooth, are often reddish in color, and the "eye" is comprised of three rather than one ocelli. Another adventive, *Trichoniscoides sarsi*, looks similar to the *Trichoniscus* spp., but each eye is just one ocellus like *Trichorhina*. Some tiny native species are similarly small, but quite rare, and may be found in certain ant nests.

ISOPODS IN DISGUISE

Cryptic coloration and hiding are the primary active methods terrestrial isopods employ to avoid being eaten. Most are gray to brownish and minimally visible among wood, dead leaves, and rocks. They have no chemical defenses to poison a predator or impart an unpleasant odor or taste (though the very rare species are colored to resemble chemically protected creatures). Running and dropping in order to conceal themselves in the closest crack and tightly gripping to these surfaces are part of the defense. Some species pull in the legs and adopt a defensive curl. When in this curl they may refuse to move for many minutes in hope the predator loses interest. It is amazing how many small predators lose interest in something as soon as it stops moving. However, the most conspicuous, famous, and fascinating defense is the ability of various unrelated species to transform into spheres.

Volvating *Armadillidium vulgare*.

The transformation from a natural creature with legs and antennae, to an unnatural globe or sphere most likely has the least to do with disguise, but does play a pivotal role in escape and evasion of predators. A round ball will be much harder to see among small rocks and gravel and many volvating species live in rocky areas. It seems unlikely as camouflage since isopods often do not stay in the curled up

The largest pill millipedes like this Madagscan *Globotherium* sp. are gargantuan compared to terrestrial pillbugs.

Adult females and nymphs of the Pillbug cockroach look like volvating isopods but adult males have wings and can fly, *Corydidarum pygmaea* from Japan.

Three-banded armadillo. © Bob Eastman

position long enough for a predator to move out of sight. Cultured specimens often refuse to stay curled up for more than a second. (I had briefly considered using glue to make one stay in a ball long enough for the camera to focus.) When a predator tries to grab a ball it can be much more difficult to hold onto. The transformation between the two states can cause a predator to lose its grip. A ball will bounce, roll down slopes, and fall into holes. A ball can also be very hard to take a bite out of unless it fits entirely within the jaws.

Conglobation and *volvation* are alternate terms used to describe the act of rolling into a ball. In this text volvation will be used because it is more precise and less cumbersome. *Conglob-* means *together* or *with globe* while *volv-* means *to roll* or *roll up*. Conglobation is often used for cuckoo wasps which roll into a glob but do not form an appendage-less perfect sphere, while volvation is common usage for pill millipedes which are much more similar in form and function to pillbugs.

Isopods are probably the first volvating creatures on earth since they are an ancient marine group whose body readily adapts to this form. There are some volvating species found in marine and littoral environments, but a transition in morphology and behavior is seen from sea to shore and further towards the deserts. Rolling into a ball is one strategy to survive "difficult" environments (e.g. arid climate). *Venezillo* spp. range widely over southwestern U.S. desert areas. The genus *Armadillidium*, the richest in the Mediterranean, is often found in dry, rocky mountain areas.

Other volvating creatures include mammals like the pangolin and armadillo (there is an armadillo lizard though its defensive coil is

Venezillo arizonicus employs physical adaptations and burrowing behavior to survive some of the driest North American biotopes, specimen from southern Arizona. © Peter Clausen

hardly a sphere) and a large assortment of unrelated invertebrates. In millipedes, three different orders have members that display this defensive behavior; keep in mind every isopod falls within a single order. Millipedes are primarily terrestrial and the oldest almost certainly rolled into a spiral, but more recent types can roll into an impregnable sphere. A number of cockroaches form hemispheres somewhat like the isopod defensive curl, but only

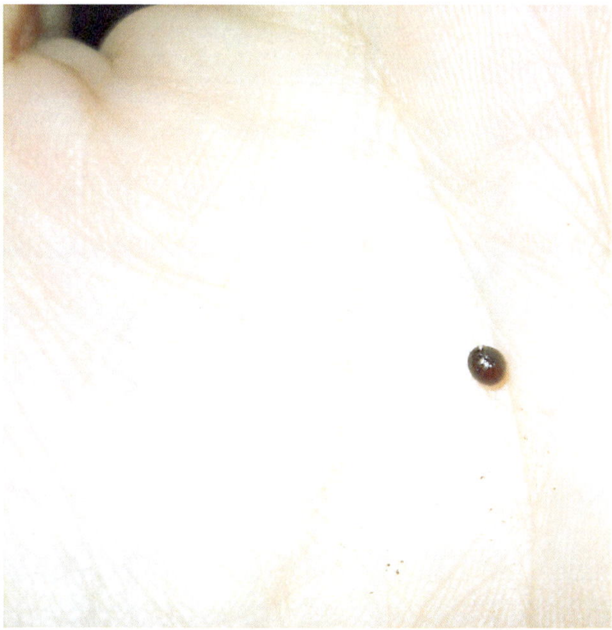

A first instar pillbug cockroach is tiny but larger than any isopod manca and gradually darkens in color after being born white. (Isopods require a few molts to gain color.)

a few rare species such as the tropical Asian pillbug cockroach, *Corydidarum pygmaea*, are able to volvate. There are also small volvating beetles known as pill scarabs. A number of fossil trilobites could fold up similarly in defense, but the end result was not terribly spherical. Lastly, some marine chitons (molluscs that resemble snails or limpets but have segmented "shells") are capable of volvating, though the response is not very rapid and takes place only after one has been peeled off a rock's surface.

Although the end result may seem the same, the way in which the body plates fold up together can be quite different among arthropods. In pill millipedes each body segment has a recessed lip and tapers towards each side so the plates come tightly together at the central sides of the ball. In isopods the sides are less tapered and fold under the segment in front of each. Pill cockroaches have similar thoracic segments,

There are species of green pillbug cockroaches not to be confused with green pill millipedes.

Pill millipede body segments have recessed edges unlike isopods, *Arthrosphaera brandti*.

Volvation of pill millipede results in no gaps between overlapping segments and disturbed animals can stay in the protective sphere for hours.

Pill millipedes not only mate differently but also have proven nearly impossible to reproduce in captivity, unlike the easily cultured pillbugs.

Some of the chitons appear capable of volvation. Unidentified tropical species that has been reproducing in my reef tank for a few years.

but three versus seven. The relatively large abdomen does not have segments that extend lower than the body, so they do not overlap when the sphere is formed. A close look at the way the edges of the segments forming the sphere overlap can reveal what animal is being depicted, but there is a much easier way to differentiate isopods from other volvating arthropods. Eight small dorsal pieces that make up the abdominal tail are visible. In millipedes and cockroaches the last segment is a single large plate. In volvating beetles the entire sphere is only three segments and parts of the legs are visible.

Despite incredible differences, the confusion of isopods with pill millipedes likely has resulted from lack of familiarity with either. While pill millipedes have 40 or more legs versus the isopods' 14, they can stay coiled for hours, so it is often impossible to count legs on a healthy pill millipede. It can be difficult to get isopods to stay in a sphere long enough to snap a photo. Increased interest in isopods meant confusion was growing less common by 2006, but then U.S. imports of giant pill millipedes (order Sphaerotheriida) were stopped along with any confusion. Of course people still come along now and then asking where to find the giant green pillbugs (Madagascan millipedes). In Europe, smaller pill millipedes from the order Glomerida are often confused because they are common and around the same size, but

Trachelipus rathkii (cc) USGS Native Bee Inventory and Monitoring Program

in the U.S. the Glomerida are practically impossible to find, even rarer than native pillbugs like *Venezillo*. In at least one book a *Glomeris marginata* photo is labeled as a crustacean (Tait et al 2005). Even nature attempts a little confusion, as at least one European pillbug species, *Armadillidium maculatum*, mimics the black and white warning coloration of the common and heavily chemically protected *G. marginata* (as the millipede produces defensive chemicals similar to Quaaludes).

COMMON NAMES

Terrestrial isopods are known by many names which may or may not have to do with their biology. The most common are pillbug, armadillo-bug, roly poly, slater, sowbug, woodlice, and "potato bug." Pillbug is a great, lasting name, but it only refers to volvating species. Armadillo-bug works likewise, but is long, not very common, and needs a hyphen to differentiate it from true bugs (some of which conglobulate). Roly poly is also widely used for volvating species, but usually only by very young children. Slater is normally reserved for the large, long-legged shore species, but it has been used for common garden varieties, at least in New Zealand (Forster and Forster 1980). Sowbug refers to isopods feeding on scraps and does not preclude volvating species. The name sowbug means "female pig"-bug and has fallen out of favor in common use because it sounds unpleasant and isopods do not look like pigs. Usually if an animal does not roll into a pill it is just an isopod. The name woodlice refers to their common behavior of clinging to wood, but lice are parasitic insects and isopods are not found on live wood, so the reference is based on ignorance, not a unique understanding of parasitic marine biota. This term is rarely used in singular because the reference is to a group. In the northeastern U.S. where I grew up, they are often called "potato bugs," but potatoes are not a crop grown commercially in my current area due to soil conditions. Isopods don't burrow deeply nor have a special affinity for buried potatoes. In the western U.S., "potato bugs" are massive, burrowing crickets.

The name *isopod* is somewhat of a misnomer when reduced to its root meaning since it comes from the words "same" and "foot." This refers to the second pair of thoracic legs ending in single claws, the "same" as posterior pairs, when this pair terminates in pincers for most other crustaceans. However, the first pair of thoracic legs is modified into tiny mouthparts so different they must not have been recognized when the order was first named.

Some common names are more correct than others. To begin, there are pillbugs, and there may be pill-bugs but there are no pill bugs. The space denotes systematic accuracy of the noun (not the descriptive verb) and isopods are not bugs. Consider: a ladybug is not a bug, nor a ladybird a bird, but a lady beetle is a beetle. This rule would apply to woodlice, sowbug, and potatobug if I were planning to use those unhelpful terms elsewhere in the text. The hyphen is considered unnecessary in codified naming schemes, but could be used in place of a contraction to display systematic inaccuracy.

The common name *pillbug* is generally straightforward but still has been debated. The name is like a snail to a slug (a slug with a shell is a snail). Simply because many marine snails are more closely related to marine slugs and certain freshwater snails while terrestrial slugs are more closely related to many freshwater snails does not mean systematics can be used to make shell-less slugs snails. Some people have tried to award taxonomic significance to pillbug and claim that many "pillbugs" cannot volvate while many "non-pillbugs" can. However, the common name refers to a physical attribute and only those isopod species that roll up into a sphere or "pill" can be called pillbugs.

2
BIOLOGY

TERRESTRIAL ISOPODS are helpful creatures that feed on decomposing plant and animal matter and help to recycle these materials back into the environment. They are considered beneficial as they break down dead plant material and provide food to lizards and birds (Werner and Floyd 1994). Their beneficial biology is the reason interstate transport has not yet been considered for regulation (lack of regulatory consideration/requirement was according to a presentation given on permitting requirements at the 2009 American Tarantula Society Conference in Arizona by W. Wehling, head of USDA permitting, and could change at any time). There are 113 types of terrestrial isopods in North America, north of Mexico, with only 2/3 of these being natives. The fact agriculture has accidentally introduced and spread 1/3 of present U.S. species, mostly from Europe, around the U.S. may play a role, since there is intrinsic difficulty in regulating accidents, introductions already occurred, and none have proven harmful. Still, if a keeper must get rid of a culture and it cannot be traded or given away, it is important to remember captive stocks should never be released and old substrate should never be discarded outdoors unless it has been thoroughly cooked. In many areas it is illegal to release non-native fauna of any type, even if it can be found elsewhere in the state. Depending on local rules, a toilet may be an option for disposal of the tiny animals since even aquatic isopods would not survive minutes if flushed. (Any imaginary sewer survivors would then be eaten by the alligators.)

The reason terrestrial isopods have been terribly slow to gain even minor pet status compared to some other arthropods is small size. The largest, old specimens of the biggest species only grow to around 20 mm. There are some relatively huge, shore-dwelling *Ligia* that can grow as large as 35 mm, but specimens of the largest species rarely grow so large, have proven difficult to keep alive, let alone breed, and are not fully terrestrial.

BIPHASIC MOLTING

Biphasic just means there are two phases. Members of the order Isopoda undergo a strange sort of molt where only the front half of the body molts; the back half is shed later after the exoskeleton has hardened. (The rear portion can molt first, but the order is immaterial to function.) The front half includes the first four pairs of thoracic legs, while the back half contains three pairs of thoracic legs. This affords them mobility during molting, since they can still run around with only six or eight legs operational. In terrestrials the front eight and back six are virtually indistinguishable, but in many aquatics like *Asellus*, the two leg groups are different in size and shape. Isopods generally cannot walk as quickly with half the legs functional, but by splitting the molt into

Molt sequence, part 1.

Molt sequence, part 2.

Mancae of the “peach pillbug.”

two phases they are able to avoid the stationary vulnerability that is usually the hallmark of having an exoskeleton that must be shed.

Molting *P. muscorum* adult. The new exoskeleton appears fully colored immediately, rather than pale or white, because the exocuticle is see-through.

Biphasic molting may be the single most important feature that allows isopods to so fully master the shoreline, especially in rocky areas where burrowing is not possible for other creatures. Even terrestrial isopods may live in areas of periodic flooding during which even a short period of restricted mobility would be deadly. Many other crustaceans like crayfish and crabs have a large carapace or cephalothorax covering the body, so there would be no benefit to molting the back half at a later point in time. Insects do not have enough legs (split two and four, two legs would offer little or no mobility) and use cell construction or hanging for protection. Myriapods have plenty of legs but dig underground burrows for molting safety. Arachnids may have enough legs but employ hanging or burrow construction to reduce the chances of being discovered by hungry animals during this vulnerable time. Lack of biphasic molting may be why none of these other arthropods have so thoroughly conquered such volatile habitats as the isopods nor crossed over notably between land and sea.

Another uncommon feature for terrestrial arthropods is that the isopod's body pigment is located beneath the exocuticle of the exoskeleton. For most insects and arachnids, an especially white animal is simply a freshly molted animal that will darken to normal coloration over the next few hours. This is not the case for isopods because their exoskeleton does not darken. A white isopod is either dead and hollow, a shed exoskeleton, a manca or small immature, or a white form. One reason dead isopods are rarely kept or offered is that both dried and liquid-preserved specimens become discolored and end up white or rotting gray no matter how pretty the live coloration was. Small size, limited variability, and fragility of the exoskeleton are other reasons.

Dead isopods begin to rapidly discolor.
© Henry Kohler

Other than the unusual biphasic aspect and exoskeletal transparency, molting is similar to other familiar arthropods. A new, softer exoskeleton develops below the old exoskeleton, which is partly dissolved at the same time from underneath. As usual, this process results in a milky or gray appearance to the exoskeleton when air gets beneath the old armor, sometimes a few days beforehand. At first the new exoskeleton is soft and walking is not possible (for the legs that just molted). Afterwards the exoskeleton hardens through a chemical process. It takes many minutes to hours for the new exoskeleton to obtain usable rigidity and up to a few days for maximum hardness. The process is known as sclerotization because it most commonly involves the protein sclerotin, but other proteins and calcium salts can be used in the hardening process. The remains of the exoskeleton are often consumed as soon as the molt is complete, though other tankmates may eat it before the molting animal has the chance.

SOCIAL BEHAVIOR

As a group the isopods range in interaction from isolated individuals to rigidly defined and sophisticated social societies. Parasitic marine species usually do not see another isopod unless they are permanently attached to a mate or just happen to be on the same host. Many other marine species are gregarious and found in groups like nearly all those on land and in freshwater. All of these live together in loosely defined groups and interact using their antennae. Most species exhibit limited or no concern for the composition of the groups. It is not unusual to find more than one species of isopod in the same aggregation on a sponge in the sea or under a log in a nearby woods. Sometimes members of the same species congregate away from other species. Specific groups could relate to mating, but aggregations can consist of all females or immatures. Groups of the same species do seem to aggregate under a particular rock or log (and migrate together to new shelters, sometimes to be replaced by a different species group) but these might be explained by dampness preference and shifts in the moisture level caused by external changes in the environment. Some of the large *Ligia* species found on rocks on the shore display a hierarchy of individuals based mostly on size and constantly tap each other's antennae. This seems apparent when they are observed but they do not cooperate in any endeavors.

Active cooperation in terrestrial species seems to be related almost entirely to moisture conservation and can range from nearly incidental to the most complex behaviors observed so far. A number of the large, non-volvating species display a behavior that can be termed as *bunching* (Masters 1975, Kneidel 1994). Under dry conditions they actively cluster together and overlap bodies. This same clustering behavior is also known to offer an effective defense in some cockroaches from seasonally dry areas, notably the common hisser, *Gromphadorhina portentosa*. Bunching can be a very effective defense against desiccation. However, which animals get the outside position and increased protection for immatures fitting underneath are incidental. Related animals likely live nearby in nature, but relation and species does not limit the behavior. The most impressive social defense against desiccation is seen in *Hemilepistus* species found in Old World deserts. Some of these have been studied for their rigid social roles created by the need to build and defend burrows in otherwise inhospitable desert areas. The burrow can only be constructed in the spring when the ground is soft enough to work. During the rest of the year one adult must stay behind to defend the home while the other forages. A monogamous male and female pair work together to build and defend their burrow with the help of their progeny later on (Linsenmair 2007). Chemical "badges" allow *H. reaumuri* to recognize members of their own family group while frass around the entry helps them recognize their own burrow. If a female encounters a small immature from a nearby family group she may even catch it and feed it to her young (Linsenmair 1974).

Ligia exotica pair.

Bunching in a group of isopods.

Unlike grasshoppers and a number of other non-predatory invertebrates that live in loose groups but cannibalize compromised individuals regularly during molts, this is density controlled for commonly cultured isopods. They do not bother each other unless severely crowded and this can be mediated by adding shelters. Still, a damaged and dying or dead animal can be rapidly consumed. Starvation usually does not change behavior unless it is combined with overcrowding. Swarming behavior is seldom seen when animals are overcrowded but well-fed. When crowded and hungry, voracious species exhibit swarming behavior and rush in at the first smell of hemolymph (invertebrate blood). Nevertheless, even when a culture is both starving and severely overcrowded, the majority of individuals molt successfully without being consumed. Even if an enclosure is busting at the seams with starving isopods, they usually do not swarm on exposed, molting cockroaches, millipedes, etc., unless those animals are compromised (a broken bleeding section present due to unrelated molting difficulties).

RIGHT, LEFT, RIGHT . . .

I watched a Japanese game show once where contestants bet on the validity of "myths." The only one I remember from the show was about isopods alternating directions each time they run into a barrier. Whether they start left or right does not matter but the next turn is always the opposite direction. It sounded far-fetched and I do not know if anyone won. However, when that little isopod ran the maze it went right, left, right, left like clockwork.

GENDER DIFFERENCES

The aquatic *Asellus* sp. are easy to sex because the males are three times the size of the females and have large uropods, but few terrestrials are so strongly dimorphic. Male slaters (*Ligia* spp.) also tend to be larger, while their front three leg pairs can be twice as thick as the other legs. Some of the *Porcellio* like *P. laevis* also have enlarged uropods on adult males, though males and females are the same size at the same age. The enlarged uropods can be three times the length of the female's and very easy to see on large, old, and wild-collected adults. However, young adults can be breeding for a year or two in captivity before they are large enough to be differentiated, since the males' long uropods only develop notably with old age.

The pleopods are male secondary sexual organs, somewhat like those of crayfish, and likewise show adaptation in shape or form according to species (McMonigle 2004). These male appendages can be the only way to differentiate some closely related species. Also, in some species the rear pereopods (legs) can have extra spines or hooks the female does not possess.

Male uropods are notably long in this *P. laevis*.

Another way to separate males and females of most species is by checking for the presence of oostegites between the legs. Even females that are not currently reproducing will have small bumps near the inner sides of the walking legs from where they are formed. For most terrestrial species it is difficult to tell without unrolling or spreading out the animal when the female is not holding eggs or young. If she is holding, very little pressure will pop the

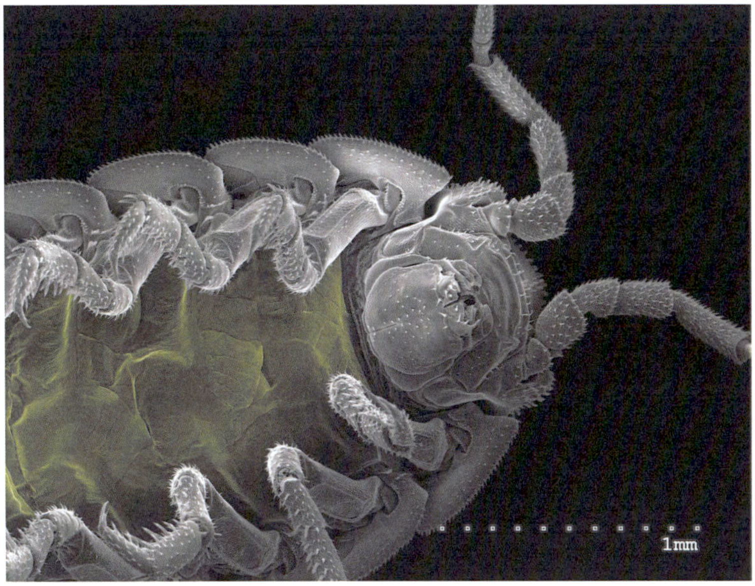

Marsupium SEM *Trichorhina tomentosa.*

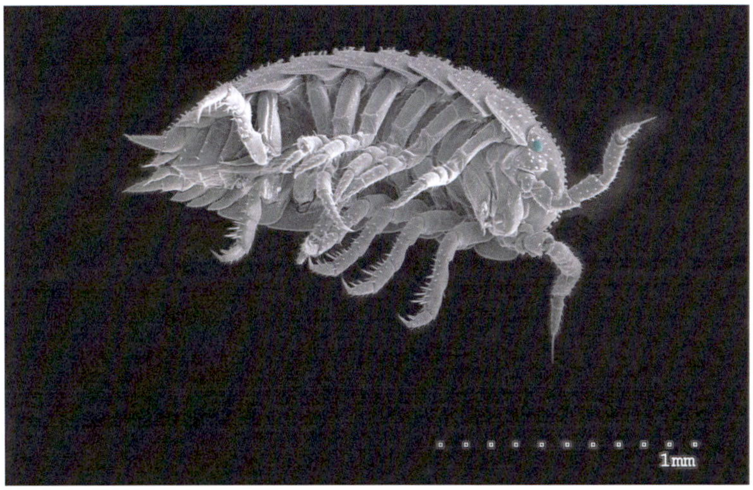

Male pleopods jungle micropod.

Dorsal body structure SEM, *T. tomentosa.*

oostegites and release eggs or young. Eggs released will simply dry out and die. Another problem with sexing is the keeper can easily damage or kill specimens trying to pick them up, let alone forcibly unrolling body parts for gender determination. If a specimen is near a molt it will probably be killed before the condition is noticed. When there is such a small number that sexing is desired, it is not worth killing a single male or female (and whatever sex there is plenty of will never be the one that accidentally gets killed). Once a keeper becomes familiar, sexing will become second nature and chances of damaging a specimen greatly diminish; yet gender determination is purely academic since there is usually no benefit to risking the death of anything in short supply. Males and females are all kept together safely.

Remember that for many species, females greatly outnumber the males. A disproportionately high number of females are normal for wild-caught animals, whether attributed to microbes, differences in natural male survival, or gender-biased trapping methods (Scott et al. 2001). A notable female bias extends to many captive terrestrial stocks, but sometimes the bias is reversed, and some instances might relate to gender switching. Gender switching is documented for the marine suborders Anthuridae and Sphaeromatidae (Allsop 2003) but I have not seen anything on terrestrial species. However, I have seen reproductive female *Cylisticus convexus* and *Porcellio dilatatus* discontinue production of young and take on seemingly male characteristics as they grew larger and obtained two and four years of age respectively.

REPRODUCTION

Fertilization is direct as males transfer a spermatophore to the female. Males usually sit directly on top of the female, not riding off the back like turtles or most beetles. This is commonly noted on aquatic *Asellus* spp., since males may hang on for hours or days. Mating is a rarely observed event for most terrestrial species, but large *Porcellio laevis* males are often seen guarding or mating with a female. Some species reproduce through parthenogenesis—that is, there are no males and each female forms eggs that develop without fertilization. This is how *Trichorhina tomentosa* normally reproduce. Other species can have a large proportion of females to males, but this does not mean parthenogenesis is occurring.

Eggs are visible within marsupium, *Ligia exotica*.

Mature females produce a number of large eggs that will be placed inside the marsupium between the legs. The marsupium will be formed by plates called oostegites that protect the eggs and allow the females to carry them around. The female keeps the eggs safe and carries them everywhere she goes, so terrestrial isopods are always portable (remember biphasic molting keeps young mobile as they grow). Oostegites can appear like a clear sac, but they are hardened overlapping plates that are easily popped by minimal pressure. If a female is grabbed, the eggs or mancae may start oozing out. Eggs released prematurely will not be put back by the female and they will die.

The number of eggs is determined by the species and the size of the female. Larger females tend to produce a greater number of eggs, sometimes more than double that of a small female, rather than larger eggs. An older, larger female can carry as many as twice the number of eggs of a small female of the same species. However, the range is determined mostly by species, as "Jungle micropods" often produce fewer than six, while *Armadillidium nasatum* can produce close to a hundred. (Half or twice either number is not going to bring them close together.) Still the micropods can be much more productive because they reproduce continually, grow rapidly, and mature quickly. One of the largest species and one with very large broods, *A. vulgare*, can produce young less than once a year even when conditions are good.

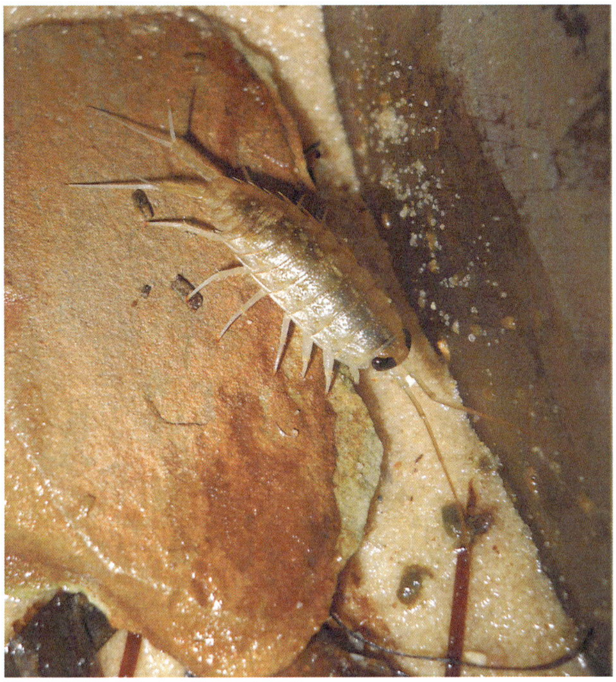

Biramous uropods are notable on some species. *Ligia exotica* from Florida.

Eggs hatch after three weeks to three months according to species and temperature. The hatchling or first instar is called a manca. Manca is only the name for 1st instar pericaridians since this stage has six rather than seven thoracic segments. This is difficult to see even under magnification since mancae are tiny, pale, and delicate. Fortunately this type of naming is rare or the crustacean worker would have to memorize thousands of names to cover every minor morphological difference, as there could be two dozen names to remember for one tadpole shrimp (*Triops* spp). Unlike most marine crustaceans that have strange-looking planktonic nauplii, the 1st instar marine, freshwater, and terrestrial isopods normally look very much like miniature adults, except they are pale in color. They will stay pale for the first two or more instars and may not obtain full coloration until maturity. Some species like *A. vulgare* and *L. pallasii* display unique juvenile coloration that is not a simple transition between pale young and adult patterns.

LIFE CYCLE

Isopods are slow growing compared to feeder insects, such as fruit flies that take about ten days from egg to adult or crickets that require two months. Some of the fastest, like "jungle micropods," take nearly three months, while *Porcellio* species must develop for six to nine months. Even within sibling groups, animals reared in the same enclosure can mature a few months apart. An *Armadillidium* female can take more than a year and a half before she is ready to produce her first brood. Unlike fruit flies and crickets that die a few weeks or so after laying eggs, sexually mature isopods continue on for years and continue to reproduce and grow. Differences in environmental conditions such as food and temperature, as well as differences between species and even individuals, can have a huge impact on the time spent in adulthood. However, most terrestrial species live two to three years after reaching maturity. The freshwater *Asellus* never seem to live even a year after maturity, while it is likely the incredibly variable marine species range in adult longevity between weeks and decades.

3
CLEAN-UP CREWS

THE HISTORY of keeping isopods is the history of the clean-up crew (also seen as clean-up-crew, clean up crew, or cleanup crew). Isopods have become popular in recent years almost entirely for their use in keeping other pets healthy. The oldest text reference with an explanation and the term "clean-up crew" is probably the reference in *Giant Centipedes: The Enthusiast's Handbook* (2003), though it goes as far back as the late 1990s when I first began trading the isolated "Spanish orange isopods" and "giant springtails" with dart frog enthusiasts. The idea came from the clean-up crews commonly offered for use in the mini-reef aquarium. The concept is similar, though isopods are mostly used for control of rotting, excess, scattered food and prey materials that otherwise result in contaminated air and harmful mite and pest infestations. Marine clean-up crews are composed of blue-legged hermit crabs that eat hair algae, and turbo snails which feed on slime algae, though there are the less commonly used emerald crabs that target *Valonia* spp. algae, and peppermint shrimp that eat *Aiptasia* anemones. Terrestrial isopods are superior to reef clean-up crews since they can be easily reared in captivity, can be shared among enclosures, and rarely need to be replaced. (Turbo snails, blue-legged hermits, etc., are not realistically reared in captivity and have to purchased at regular intervals as they die.) Moreso, isopods can be employed as feeders, as a clean-up crew, or a combination of the two. Like reef clean-up crews, they add some life to the vivarium and help establish balance in small artificial ecosystems.

Uses are varied. Most isopod species are employed to keep damp terraria clean of rotting prey and prey parts, certain fungi, and rapidly decomposing plant matter. Low humidity is not a reasonable alternative to control pests for many creatures since it kills the main inhabitant as well. Isopods are most often put to work in amphibian terraria, whipspider and humid tarantula cages, large forest scorpion enclosures, cockroach containers, assassin bug terraria, and centipede jars. Most predators will ignore isopods and allow them to do their work unless they are not sized correctly or the keeper forgets to feed the predator for a while and it grows extremely hungry. Large centipedes and small tarantulas are especially prone to taking out every last janitor if better prey has not been offered recently.

The items isopods feed on can poison the air in the cage or lead to infestations of grain mites, phorid flies, fruit flies, or other destructive and annoying cage pests. They eat rotting items and prey pieces the main inhabitants either are too large to find or are too picky to eat. Many large predators discard prey parts or a large pellet of incompletely cleaned chitin. Without a clean-up crew (even with manual removal) volatile materials often results in

Nature`s clean-up crew, *A. nasatum* and *A. vulgare* turning a dead tulip plant back into soil.

Whipscorpions feed on large and small isopods so even micropods do not necessarily make a clean-up crew capable of surviving their voracity.

grain mite infestations that stress or kill most arthropods when numerous enough. Isopods will also prevent certain types of entomophagus fungus and phorid flies that feed on decaying arthropods by removing the source of food.

The most commonly encountered grain mite species is probably *Acarus siro*, as it infests mealworm media, fruit fly media, and cricket cultures, as well as amphibian, arachnid, and centipede enclosures. Full-grown reproductives are much smaller than a grain of salt and engorged adults are seldom bigger than the period at the end of this sentence. The hypopus (dispersal) stage follows an outbreak as food runs out. The mites molt into suction cup-like forms that tightly adhere to other invertebrate exoskeletons and can block breathing holes and lead to death if outbreaks are severe enough. These ubiquitous mites are unsightly and sometimes result in allergic reactions. Large numbers will make skin crawl even for those who are not allergic. If large

Entomophagus fungus that grows on dead invertebrates and may infect healthy animals in proximity can be prevented with clean-up crews.

numbers of tiny white mites are seen, the substrate should be replaced and the cage cleaned. This is most helpful when done before mites have begun molting into the hypopus stage. The final step, beyond employment of clean-up crews, is to remove dead prey remains and reduce feeding. Grain mites also infest most rotting fruit and processed animal food pellets, and isopods clean up these scraps.

Isopods are most commonly used in dart frog terraria for clean up and prey.

Live plants are ignored, while leaves become fair game as soon as decay sets in. Isopods consume the decomposing fruiting bodies of mushrooms, but do not feed on most live wood or leaf fungi. Fortunately, the fungi they do not eat are harmless to other invertebrates. Excess fruit or food jelly can be eaten well before fruit fly maggots have a chance to grow.

Isopod clean-up crews are incredibly useful and helpful, but they are not magical. They simply outcompete enclosure pests and if the population is too low, they will not work efficiently. Large volumes of rotting materials should still be removed manually by the keeper. Isopods can be expected to clean up the residue, not massive quantities of rotting items. In order for a clean-up crew to work, it should be chosen for the appropriate size and there must be enough to consume all rotting materials. If too many are added the excess will die out. The exact number will vary greatly by how much food is being eaten, how much is being fed, various aspects of the enclosure, and the type of isopod. Fast breeders can build up to good numbers after a few months and few may be needed. If they also serve as food for the cage inhabitants, the culture container should be separate.

It is important to choose the right janitor. Some types will be poorly suited or quickly eaten. Micropods are among the best clean-up crews because they are least likely to be eaten and they reproduce rapidly. Micropods do well in hot, wet terraria, but can be impeded by tiny dry barriers or can reproduce so much they need to be manually thinned out a few times a year. Dwarfs and medium-sized species like *Porcellionides* are a sort of halfway point between the micropods and large, interesting species. The orange *Porcellio* are commonly used because they are useful cleaners, are easy to control, and survive a greater range of environments, but also provide a little food and entertainment. Cryptically colored and species more prone to burrowing are employed when the singular goal is an invisible clean-up crew. Pillbugs and the larger *Porcellio* are generally suited to less damp and cooler terraria and almost never outgrow their welcome. *Oniscus* and *Philoscia* are more limited and generally used when the goal is zero reproduction of the clean-up crew. However, *Porcellio* and *Armadillidium* are hardier and usually will reproduce little, if at all, if rotten leaves are not included in the enclosure. Chapter six lists stocks that are commonly available as clean-up crews and addresses some further considerations.

THERE IS A DARK SIDE

Isopods should not be used with beetle larvae, millipedes, or anything that directly competes with them for food. They might even outcompete less virile, rare, roach species, but since the latter are finicky it would be tough to prove with hard data. In small numbers isopods are harmless to everything except exposed beetle pupae, but when they number in the hundreds and thousands they can eat soft items that do not move. They usually will not bother millipede eggs in sealed capsules but any cracked or broken capsules will be emptied by hungry maws. Molting animals can also be at risk but problems are rare because the movement of a molting arthropod will scare off the isopods and many hang out of reach of the isopods or have protective mats of hair. Reports of isopods attacking molting animals are not common, and probably the result of a sick animal dying during the molt and then being consumed. If factual, such problems only occur when the isopod population is ridiculously high in the cage. Isopods are not predatory but they do take advantage of nonmotile available food. If they did not readily feed on dead invertebrates they would be worthless as a clean-up crew.

Large numbers of starving isopods will exhibit swarming behavior. Hungry animals are attracted to feeding frenzies and they do smell hemolymph. My whipspider cages remind me of bat caves because as soon as a ball of chewed chitin is dropped the isopods swarm and completely cover and tear it apart. This also happens to a half eaten cricket. If a whipspider were to fall while molting it seems it might suffer the same fate, though I have not yet known one to disappear this way. (If it did fall it would die by the next day anyway. Its imminent death would only be 24-36 hours more rapid, not more certain, and reduce time suffering.)

A minor negative aspect that may go unseen for months or years is that huge numbers will eventually chew up almost any cage decoration. Logs, both hardwood and cedar, and cork bark can be chewed to nothing over years. Even compressed hardboard and some types of rocks will show wear over time.

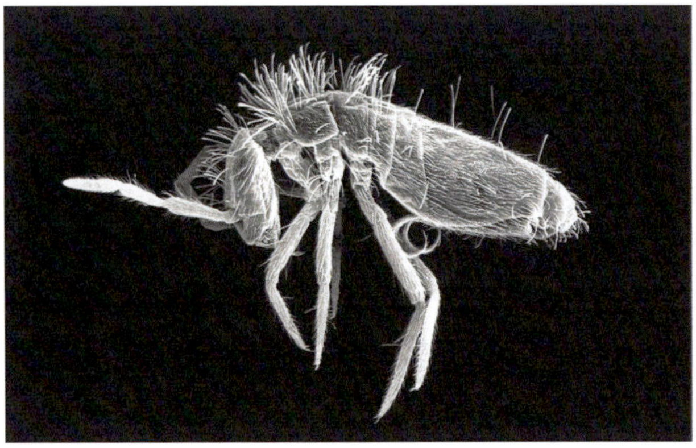

SEM of "giant" springtail with furcula in normal position before a jump.

SEM of "giant" springtail with furcula in released position.

COMPARE AND CONTRAST: SPRINGTAILS

A discussion of clean-up crews would come up short if springtails were not considered. The singular complement or competition for isopods are springtails, and presently only a single species, the "giant" springtail, has proven of any real function. It is easy to raise all sorts of springtails, but the vast majority are just too small or simply not prolific enough to function as a clean-up crew. Isopods versus springtails is not an either/or proposition, they are often used together.

Springtails are useful in damp enclosures with millipedes and beetles, because they cannot harm or unearth vulnerable stages. They may also be needed if the cage inhabitant readily eats small isopods. The biggest drawback to springtails is the cage can look rather infested. The visitor almost always asks, "What are those things all over?" Isopods stay relatively hidden and even huge numbers can remain unseen. Many of the isopods hide during the day even if a colorful form is chosen, whereas none of the cultured springtails are cryptic in color and springtails climb and jump all over the cage. Despite springtail infestations that look like an introduced Australian wildlife documentary, they do not harm, stress, or eat other arthropods.

Close-up shot of "giant" springtail. © Henry Kohler

A group of the commonly cultured entomybryid springtails popularly known as the "giant" springtail.

VERSUS PREDATORY MITES

Predatory mites are beneficial helpers in certain circumstances, but in a terrarium where food is in supply to keep them going, a large culture of these mites is harmful to some creatures. I have found them functional for control of grain mite infestations such as with infested, damp rhinoceros beetle substrate, and dry substrate for grain beetles and dermestids. In the case of grain substrate, tossing out and replacing the media is not a real option since all the mites cannot be completely eradicated and new media will be rapidly reinfested. It usually takes three weeks to a number of months, but the more massive infestations of grain mites are usually the most rapidly controlled. Even

This predatory mite is *Hypoaspis miles*. It is larger than most grain mites but is only about 1/50th of an inch (0.5mm). It lives in the upper layer of damp substrate, often on the surface. © Henry Kohler

hypopus stage grain mites are consumed. They often completely eradicate grain mites over time and then die out themselves. Predatory mites can eradicate nematodes and reduce fungus gnats, but it takes months and reduced or absent food. These would be superior to isopods in most applications where the singular goal is grain mite control, except they do not work in many large or heavily populated cages. Predatory mites can work on various media

Reduction of problems such as this grain mite infestation are a primary reason for using clean-up crews.

infestations of different moisture levels, but when the source of the problem is dead arthropod bodies or parts of any volume, they have very little impact on pest populations. Predatory mites can be expensive and are difficult to culture except in those enclosures where they are unwanted. I know at least one major vendor who refuses to carry them due to the uncertain impact on a variety of invertebrates. Their presence severely depresses and can lead to eventual failure of springtail and pillbug cockroach cultures. There is no hypopus stage. Though large compared to grain mites, these are very tiny and do not negatively affect aesthetics.

4
ISOPODS AS FEEDERS

Terrestrial isopods are excellent prey items for a number of commonly kept predatory creatures. They may be used as the primary prey for some creatures like aquatic true bugs and vampire crabs, but most often are used as a supplement to introduce variety in the diet. Even the most rapidly developing species will require large starter cultures or months and years of build up to provide significant production. Considering the great number of adventive species that can be collected around rocks and logs piles in most back yards, wild-collected animals may be an option, but native species should be avoided. The likelihood of introducing harmful residual pesticides, bacteria, or viruses should be weighed against the numbers and die-off of the predator being fed. Pillbug species are least commonly used as prey since they present a feeding challenge to some predators, while the most commonly available *Armadillidium* species develop too slowly for most uses.

There are a few options for gathering wild adventives. Beyond locating rocks and logs where isopods might be hiding during the day, the keeper can place rocks, logs, or boards in areas for easy access. Of course, shelters should be as far as possible from land where pesticides may be in use. Even in very cold areas, isopods can be collected from under strategically placed shelters in the dead of winter, unless the snow is too deep to locate the shelter. A step towards farming rather than trapping would be to dig out a shallow area and arrange crumpled paper under the board or wood. Food in the form of leaves and table scraps can be placed on top of the paper to discourage worms.

This potato trap caught four *A. nasatum* the first night.

A potato trap can possibly collect large numbers of animals during warm periods. The potato trap is mentioned in a number of scientific and research papers in the 1980s and 1990s, while the oldest reference in text I have encountered is from *The Encyclopedia of Live Foods* (1975). A hole is bored part way through the potato (or all the way through and a chunk replaced to seal one end). An apple corer or 1/2" copper pipe can be used to make the hole. The hole offers food and a long, moderately narrow, moist area to entice and keep the

Giant water bugs feed on isopods and leave an empty, colorless shell behind.

This F2 subadult vampire crab grew much more rapidly than control siblings when isopods were included in the diet. Early instars were fed *T. tomentosa* and later *Porcellio* sp.

This male veiled chameleon ate a number of isopods when it was very small.

Herbert`s giant water bug, *Abedus herberti*, eating an isopod.

isopods. The trap is left out overnight and the contents simply shaken out into a waiting container and harvested as needed. If the hole is round and cut all the way through, it can be easier to empty the trap. The trap is best set near a woodpile or area with large rocks.

I only recently noticed potato trap references and made a potato trap to try out. I placed it next to a woodpile filled with isopods. The next day I found four *A. nastatum* inside. I repositioned it and caught only one tiny *A. nasatum* the next day and nothing the following day. I moved it to a wooded area prone to seasonal flooding with healthy isopod populations and at least five species. Over a week the trap caught a dozen *Trachelipus*, half as many *Oniscus,* and a couple more *A. nasatum*. During this time it was positioned against objects, oriented differently each night, and a few different types of holes were dug into the potato in hopes of attracting or retaining more animals. I could have caught ten times as many isopods had I used the same time to flip rocks or wood. Still, the trap can last for months despite holes being chewed in it.

Isopods are the only widespread terrestrial arthropod used as a feeder that has a significant portion of the chitinous exoskeleton impregnated with calcium salts. Depending on the predators ingestion methodology, this calcium is likely more digestible than lime powders or solid bones of small vertebrate prey. How the calcium in the exoskeleton of isopods is absorbed by predators is uncertain, but isopods have a higher percentage of calcium salts in the exoskeleton than other terrestrial invertebrates. Crickets, mealworms, fruit flies, cockroaches, etc., use almost entirely proteins to strengthen the chitin. Only certain millipedes (specifically those from the order Glomerida or "small" pill millipedes) have more than a trace and they are among the most chemically protected and unpalatable of invertebrates.

The most common use for isopods as prey items today is feeding dart frogs and salamanders. Older texts mention their use in feeding newts and frogs (Masters 1975) and toads (Kneidel 1994). Terrestrial isopods are an excellent supplemental prey for lizards such as young chameleons, but isopods dry up rapidly when exposed and most lizards will not discover them unless they are offered in an empty food dish. If they are not eaten within an hour or so, they die and will never be eaten.

Predatory insects, like *Platymeris* assassin bugs, waterscorpions, and giant water bugs, do well on mostly or entirely isopod prey. The reason they are so useful for aquatic bugs is not spectacular nutritional value nor available calcium, since the exoskeleton is not eaten. Rather, they work well because they are lower in fat content and almost never foul the water like crickets or cockroaches. Another positive for aquatic predators is that isopods can walk around under water, sometimes for hours, without drowning (while a cricket or mealworm can drown in seconds and the predator often loses all interest). I have used *Porcellio* to rear *Calosoma* ground beetle larvae as the grubs are surprisingly timid and find other prey too difficult. Praying mantids will eat them, but usually the isopods just dry up on the bottom of the cage and are never eaten.

¡Yo quiero Isopoda!

Even voracious mantids like this one rarely finish the whole isopod because they have a hard time chewing it up.

Isopod hunter eating a *Porcellio* sp.

Isopod hunters, *Dysdera crocata*, are among the most widespread and common of spiders that specialize in eating isopods, pair in courtship.

Other common pet crustaceans do better when fed isopods versus other available prey. Crayfish love to eat them and might grow better when they are included in the diet. I have reared vampire crabs (*Geosesarma* sp.) for years and only recently discovered the early instars grow faster and are healthier when isopods are a major part of the diet. I had used a combination of pelleted fish food, fruit flies, and small crickets, but when I began adding isopods to one group of young, the improvement across the entire group compared to the control was notable. The unique benefits of reduced water fouling and resistance to drowning make them easy to use.

Isopods can be used as prey for many predatory arachnids and centipedes, but are more often used in keeping cages clean. They can be eaten by most spiders, though they have probably never been tested for singular and extensive use as prey with tarantulas due to size. Some spiders feed almost exclusively on isopods, such as the isopod hunter *Drysdera crocata*. Vinegaroons and tarantulas will eat them, scorpions and centipedes can eat them, while whipspiders generally won't touch them.

Isopods only reproduce well when a large number of adult females are kept, because they take six to twelve months to be of reproductive age, versus as little as a week for fruit flies and eight weeks for crickets. Cockroach prey may take as long to reproduce, but a single reproductive *Blaberus* spp. is equivalent to the mass of hundreds of early mature isopods of the largest species.

Brachypelma albopilosum feasting on an isopod.

5
CULTURING ISOPODS

Isopods are among the weakest and mightiest of arthropods. Individual specimens can seem weak since they are easily damaged by handling and most can die rapidly when held in the hand. (Try carrying some home from a nearby wooded area in this way.) They also rapidly die from exposure and dryness. I've seen large numbers of dead in nature, possibly due to freezing. Nevertheless, they reproduce and grow at such rates that such losses are insignificant. Common culture stocks are effortless to care for and keeping a culture for decades would require very little expenditure of time or energy. This hardiness makes them an excellent choice for culturing. The following sections will outline experiences that should reduce the chances of accidentally killing off a culture.

TRANSFERRING ISOPODS

In order to transfer animals for janitorial or feeding services, it is important not to kill them. Handling can be a noisy proposition. If you listen closely it is possible to hear the exoskeleton cracking and crunching under stress. Hand counting is easy for large, sturdy *Porcellio* and some of the pillbugs, but even these could be damaged if picked up with a finger and thumb against a solid surface. The easiest way to handle specimens is to place flat bark or pieces of wood they can climb on into the container. This wood is removed with the isopods hanging on and then tapped over the new terrarium, the hand, or a shallow container.

The counting container should be no more than a few inches high and have sloping sides or rounded corners and not deep scratches, punctured, or drilled areas. Small burs in plastic will tear the isopods apart as they are pushed up the side. For counting, animals as small as a few millimeters can be safely pulled up the side with a finger towards a waiting thumb. If too much pressure is applied, crunching noises will be heard. Damaged isopods are capable of regenerating and healing as they molt, but most often the damage is quite severe and results in rapid death.

Most isopods cling to pieces of wood. The unusual orange forms of isopods may not always be at a distinct disadvantage since rotten wood and bark are often orange in color.

The type of cage, decorations, and substrate chosen can make all the difference between life and death.

CULTURE CONTAINER

There are quite a few options for containers used to house an isopod culture. The most common are plastic tubs, plastic shoeboxes, and aquariums. Glass or plastic jars are fine, but the height is wasted space. Cardboard, wood, or even masonite boxes would not be a good choice since they are readily scaled and warp from the required humidity. Uncle Milton's Roly-Poly Playground is hard to ignore as the only cage designed and sold uniquely to house isopods, but it is a tiny children's toy and my attempts at testing it resulted in a 24 hr mortality of 50%. Originally I maintained cultures in the molded plastic small animal cages, but the excessive ventilation made it necessary to water substrate regularly. A little negligence could wipe out a culture kept for many years. There is no reason to add an unnecessary layer of difficulty so I now use cages with minimal ventilation. Whatever container is chosen, consider starting with two cages if it is a difficult to replace stock. Small differences can add up. If one culture dies out it may be easy to determine the cause through comparison.

Even pressed hardboard can be damaged over time.

Climbing of thin films and corners is something to consider in enclosure planning even though isopods cannot adhere to smooth surfaces.

Climbing is not usually a consideration since isopods have no adhesive pads that allow climbing, found on many other familiar invertebrates. Also, the common behavior of hiding may conceal climbing tendencies. Isopods are lightweight and can climb up the silicone bead in the corner of an aquarium. Small individuals and species readily climb up nearly invisible films of dirt on smooth container walls. Even a thin layer of condensation can sometimes provide tiny animals a means to climb very tall, smooth surfaces.

Moisture, ventilation, exposure, airflow, and humidity are important, interrelated considerations for choosing the culture container and lid. Humidity and moisture are terms often used interchangeably, but humidity is the amount of water vapor in the air, while moisture refers to the amount of water or water droplets in the substrate and on the glass. High water vapor levels may reduce moisture loss, but animals cannot drink it. Airflow increases water loss even in the presence of high humidity. I have seen greenhouses with 100% humidity that will dry out an exposed isopod within minutes. Substrate materials or wood can shield animals from water loss due to lack of humidity or excessive airflow. The need to retain moisture (rather than avoid predators) is most likely the greatest reason terrestrial isopods hide under flat objects or burrow.

Semi-aquatic shore species can be given a bowl of saltwater but a slanted, porous rock must be added to let them climb in and out.

Moisture gradients caused by different temperatures on opposite sides of a container can cause dangerous conditions where some animals become stranded and dry out though the other side is dripping wet and overall humidity high. This is common on shelves facing a wall. Water vapor condenses on the cold side, so over time the warm side can become bone dry and the cold side dripping wet even with severely restricted ventilation. The container can be rotated occasionally or the substrate mixed by hand every few weeks.

With all medium to small species, almost no ventilation is required. A solid glass lid is placed on the cage or a plastic shoe box is used with no extra ventilation. A few thumbtack holes are made in the lid of small plastic containers. For larger species like *A. vulgare* and *P. laevis* it is best to drill half a dozen or more 1/16″ holes in the lid of a plastic tub or bucket. The determining factor is not always size, *A. nasatum* can average the same size as *A. vulgare*, but are not sensitive to limited ventilation. For *Ligia pallasii* that require some areas of pooling water, I used a screen lid on a 5-gallon bucket. The ambient humidity, airflow, temperature, number of animals, food, and moisture of the substrate are all considerations. If adjustment of the moisture is required more than once every few months there is too much ventilation. If moisture builds up and animals seem to be dying for no reason, there may be too little. The amount of air in the container also relates to ventilation since more open space reduces the need for ventilation.

SUBSTRATE

The bottom of the cage should have substrate for moisture retention, and since isopods cannot grab onto smooth surfaces. It would be very easy to dry out isopods in a container with no substrate even if there were no ventilation. I have used sand, gravel, potting soil, composted manure, beetle frass, coconut fiber, and peat with good results. Some substrates prove overly sticky and discolor or slow down the animals too much. Others can lead to excessive mold growth over time depending on the conditions present (humidity, moisture, food, and temperature). Specifics are why a substrate that works well for one keeper may offer problems to another. Mold may take many months

Dead leaves are a very important impetus for reproduction.

to show up, but substrate should be changed if there is a lot of mold growth. Molds may not directly infect them, but the isopods can become stuck in the fruiting bodies and filamentous growth. Sand and gravel are inert, but generally are used just for shore species when very wet areas are needed with standing water (since they do not suck up or wick moisture). Other substrates hold moisture and work equally well. Usually a substrate level greater than an inch is unnecessary.

Some researchers use the isopods' own frass pellets and a solid porous base, which may prevent some substrate problems. A 2 cm layer of plaster of Paris can be applied to the bottom of the culture container. It is cheap and easy to apply. The plaster can be mixed with biochar or activated carbon to retard fungal growth. This material retains moisture and absorbs excess water. The surface is rough and chemically harmless to isopods. Once it hardens, leaf litter is placed on top. The isopods quickly create a layer of organic matter from frass and leaf fragments into which they can burrow (pers. comm. Vilisics).

SHELTERS

Wood and cardboard pieces are commonly used for shelter. Other materials are either too heavy, too smooth, or too easily chewed up. Heavy materials will smash and kill a good number of creatures. Many isopods like to grip the underside of rough, flat objects. They prefer to have the body between a narrow space or held tightly against a surface. Corrugated cardboard is not usually used since small animals will get into the folds and become very difficult to remove safely. In some enclosures the cardboard can grow very moldy. In the case of wooden shelters, if removal of animals may be desired in the future for feeding or transplant, it is best to avoid a lot of cracks and holes for hiding. If it is necessary to remove an animal from a hole in the bark, the shelter may have to be broken apart and some isopods will

Note the extensive feeding damage to this cedar half log over a number of years.

be smashed and killed. Some types of bark have tapering ridges that offer maximum shelter and still allow the animals to be easily dislodged by smacking the opposite side with the palm of the hand. Most materials will be chewed on by the isopods and the shelter will become thinner and thinner over time. Soft rotten wood can be replaced regularly (or avoided since it can be chewed into nothing rather quickly). Despite its softness, commercially available cork bark is not consumed quickly. Tree bark from various trees can be used. I have used pine and cedar in a few cages and it is chewed up like hardwood.

Shelters have an effect on burrowing behavior. Some species that do not normally burrow may attempt to burrow in the absence of shelters. Also, large substrate particles like aspen shavings or leaf pieces can act as tiny shelters since they are bigger than the isopods. Some species that normally would not burrow in fine substrates may be found as deep as large air gaps and large particles are.

Cuttlebone can make a good shelter.

TEMPERATURE

A broad range of temperatures is generally accepted and even the tropical species can breed quite readily as low as 65° F. Of course there is a bottom temperature for each species where reproduction ceases, but this is below the bottom range of most heated structures. Temperate species stop breeding but will walk around, and some even eat, at 40° F. Temperatures far above room temperature (72°-77° F) are not necessary and can harm some species. Warm temperatures may improve reproduction but Mocquard et al. (1980) showed high temperatures only marginally accelerate reproductive onset, not its duration. Seasonal studies of isopod reproduction are not applicable to captive situations unless cyclical temperatures are used to induce seasonal reproductive periods for academic purposes or in hopes of possibly inducing difficult-to-breed species.

Other than the commonly available dwarfs and micros (listed later) that die rapidly at cold temperatures, most can be safely kept in the refrigerator, but there is no reason for it. Various insect feeders are refrigerated for the purpose of delaying development or prolonging useful life, but it does not prove useful for isopods. Rapid shifts may be harmful and temperate species can experience high losses during natural winters. (A number of dead animals are sometimes seen under logs in the field.)

DAYLIGHT

Lighting does not play an important role in captive care of most species. I use mostly semi-opaque rearing containers placed on shelves far from the closest light source. I have also kept some in a drawer. While light leaking in these containers may still afford them some sort of day-night cycle, even within a closed drawer, lighting seems to be of little importance. Isopods do not like light and this aversion to light is easily tested by giving them a choice. They will feed during the day on exposed surfaces in reduced light and when food is placed beneath a piece of bark, but usually avoid daytime feeding in exposed, well-lit areas. However, isopods do not eat all day long like a caterpillar, so a few hours more or less would probably have no effect. Since isopods are most active at night it seems extended daylight would only reduce active periods and be likely to slow down development. Many of our native species are found only in certain caves or cave systems (Jass and Klausmeier 2001) where light is probably not a major consideration.

Bhella et al. (2006) state they extended the photoperiod to 16 hrs to enhance reproduction, though the study had nothing to do with testing photoperiod results and they did not refer to a source for this idea. They likely did this because previous research on *A. vulgare* showed that lengthening photoperiods were an impetus for reproduction and extended the reproductive period (Souty-Grosset et al. 1994). There are no seasonal variations required for commonly reared captive populations and no cycle is required even for *A. vulgare* to reproduce in captivity. However, the need for day length cycles (possibly affected by temperature cycles) may be why captive reproduction of *O. asellus* and *P. muscorum* is abnormal at best. As such, it is likely only applicable to someone very interested in one of these species, or being paid to study daylight periods. Remember the idea that photoperiod is the key to breeding these two species is pure conjecture—the impetus could be temperature, diet, some other factor, or combination of factors.

Most lepidopteran and phasmid frass can be used as fodder.

FOOD

The primary recommended food are brown, old leaves that have been on the ground at least a few months. I have used leaves from various hardwood trees (oak, maple, beech, birch, aspen, ash, elm, locust, redbud, sweetgum, walnut, persimmon, plum, pear, etc.) and to a far lesser extent shrubs (various *Viburnum* spp., lilac, raspberry, rose, etc.) and grasses.

Potatoes and a few other foods can be hollowed out and used as food and shelter.

The Indian almond (*Terminalia catappa*) leaves sold for terrarium use will work but can be expensive. Some leaves are tougher and less readily eaten, but lead to strong growth, such as oak, while others like maple are more easily chewed but nutritionally deficient, so a mix of leaves works best. Harder leaves can be ground up to make feeding easier. Weathered old leaves are preferred since most fresh leaves will not be eaten until they begin to decompose. Also, excess nutrients in fresh leaves can feed grain mite problems. Fresh romaine lettuce is a readily eaten treat but I have not used it regularly nor tested how it compares or contrasts with various leaves. Rabbit or gerbil pellets can be used since they are made of grasses, but they are made from fresh plants (decay is primarily arrested when kept dry) and can invite pests and mold.

Other common foods include fruit scraps, fish flakes, dry dog or cat food, and various vegetables like carrots and potato slices. Melon rinds are refuse in most households during the summer and pumpkin in the fall and winter (an uncut pumpkin can stay fresh for six months). Usually, rather inexpensive or free food is used since isopods are not picky. It is important not to offer too much dog food or fruits as even a large culture will only eat so much. Excessive rotting food can harm the animals and cause pest infestations. In the case of any foods that rapidly decompose, only what will be consumed in three or four days should be offered and only every other week. Pet foods that sit in the cage for weeks and become engulfed in mold (if invertebrate pests have not found them first) may eventually be eaten but even more slowly. Melon rinds work well but also have a short usable life. Carrots and potatoes often do not rot for months if cut into large pieces. Most species tend to lose interest in potatoes after a few weeks. Isopods often chew caverns into large carrot, potato, or pumpkin pieces and use the food as a shelter instead. Still, keep an eye out as these can suddenly decompose, especially in hot and wet terrariums.

Algae, mostly hair algae from freshwater aquariums, is highly palatable, though it is initially swarmed and later forgotten. This is likely because the isopods are attracted to the nutritious diatoms that grow on the surface and what's left is not of much value, rather than boredom with a certain food. Other freshwater algae can be used, but not blue-green. Marine algae seems less useful, especially macroalgae like *Caulerpa* spp. Sheets of algae and prophase kelp can be purchased rather cheaply from Asian markets and can be used in place

Venezillo parvus with freshwater hair algae.

If rotten hardwood leaves are difficult to acquire, kelp and algae sheets can be useful and inexpensive foods.

of more volatile foods for easy care. Reproduction is enhanced when kelp is offered (if leaves are not available) but some of the oldest adults in the culture will fall over dead a few days later. Isopods enjoy eating Java moss almost as much as crayfish do, but it may not provide any nutrients.

Rotten wood can be a useful food since it can be left in the cage indefinitely, but will result in minimal growth and reproduction (sometimes even eventual starvation) without other foods. Wood that has not rotted, such as shredded aspen bedding for vertebrates, is of some value as food since it decomposes in the presence of moisture. It is not highly nutritious, but a handful can be added so there is something to chew on when they have not been fed in a while or are tired of the normal fares.

Since they have a measurable percentage of calcium in the exoskeleton, chalk, cuttlebone, and eggshells are often suggested as part of the diet. I never tried any of these prior to 2013 so obviously they are not at all necessary. (My *A. vulgare* culture has gone 18 years without such items.) I recently tested eggshells, chalk, cuttlebone, and lime rock with a dozen species. Eggshell and lime pieces that were thrown into large cultures for months showed no signs of wear. Isopods do readily eat the film off eggshells when first introduced, but feeding on the shell itself is imperceptible. In some enclosures I have accidentally added freshwater snail shells with the algae and these thin-walled shells have remained fully intact for years. After a few months the chalk (from a chalk board) showed very tiny areas of wear that looked like nearly microscopic rows of bite marks a fraction of a millimeter in depth and 3 mm x 5 mm around (volume less than a square millimeter in total). Some direct feeding was observed on the cuttlebone when first placed in with *Armadillidium maculatum*. Chunks placed in with some different *Porcellio* species, *Trachelipus*, and *Ligia* showed little, if any wear. Cuttlebone pieces placed in with *A. maculatum, A. nasatum,* and *A. vulgare* showed somewhat large areas of wear—as great as 3 mm deep and 20 mm x 6 mm after one month. This sounds significant, but cuttlebone is mostly air by volume. Also, cuttlebone is about 10% organic matter so the calcium might not be the attractive factor (as egg shells and snail shells in the same cages show no wear). It does not hurt to add calcium sources, but after multiple tests over five months I have seen no correlation with increased development nor improved reproduction. (Improvements can often be seen within weeks when quality rotten leaves or freshwater hair algae are kept in supply.)

Most of the larger species only produce young once or twice a year in nature and can require certain foods as an impetus for reproduction (McMonigle 2004). Photoperiod and temperature cycles might work as an alternative impetus (there is no evidence they can be used in place of quality food if one considers the cyclical nature of leaves in temperate areas) or for species that are not commonly kept. Dwarfs and micropods breed year round and do not seem to care what they eat within reason. For most species the impetus is rotten leaves, though fresh hair algae from a freshwater aquarium can work even better. If only fruits and dog food are provided, the adults grow and survive very well, but most species produce few to no offspring. Females not given a food impetus for breeding seem to end up diverting resources to their own growth because constantly breeding females can stay surprisingly small. This seems logical, but males kept in the same conditions also stay small and they cannot be investing nearly as much in sperm production. Overcrowding is different from other creatures in that it does not lead to stunting. Large, overcrowded cultures often end up producing large, and even some exceptionally large specimens. This may or may not relate to food competition. Cohort splitting where some individuals choose to invest in growth rather than reproduction is also possible.

CULTURE CONTAMINATION

Other than the basic housing considerations outlined, it is important to consider how to avoid or deal with contamination by springtails, mites, worms, fungus gnats, and accidental introduction of other isopods. These can be issues but are usually not the problem they are for other types of invertebrate cultures. Springtails are the only small creature often purposefully kept with isopods to reduce negative effects of overfeeding. Still, they can be extremely difficult to extract if they are unwanted. It may be necessary to start up a new culture or decrease moisture and food for long periods. The remaining invertebrates are never added on purpose since they add no value. Mites come in various forms but the most problematic are the grain mites that are the primary reason for using isopods as clean-up crews in terrariums. These can also cause problems in terribly over-fed isopod cultures. It may be necessary to switch to only wood and well-rotted leaves because freshly fallen leaves contain nutrients that can lead to mite infestations. However, I have never seen the normal pest grain mites make it to the harmful hypopus stage in an isopod culture under a wide range of circumstances. Nematodes and earthworms are often introduced with leaves and rotten wood if these items have not been thoroughly dried or cooked. Worms usually do not have much effect on the culture, but can thoroughly overrun the substrate. Dark-winged fungus gnats are annoying little flies whose see-through maggots consume wet rotten leaves with gusto. The adults like to fly into human noses and eyes. Gnats can be prevented through airtight containers with fine screen vents or greatly mediated by not mixing leaves into the substrate and not offering too much leaf litter at once. Fortunately, isopod frass does not act as a media for fungus gnat maggots (while millipede frass does).

Free living mites are sometimes found in isopod cultures. These types do not have a hypopus stage, are much larger than the common grain mites, and are usually red or black. © Henry Kohler

Possibly the most annoying culture pest is the micropod *Trichorhina tomentosa*, because it is the only thing readily capable of overrunning other isopod cultures. If these are in a terrarium or container on a shelf or stand above another culture they'll almost always show up. These can overgrow other species and are mostly annoying because they can be difficult to pick out from immatures of other species. You only have to miss one white micropod. Other pests can mostly be mediated through changes in feeding that give isopods the chance to outcompete them, but not these, since they are isopods. Nevertheless, if the main isopod is a temperate species and the culture container is not too large, it can be placed in a refrigerator overnight for an easy, rapid solution.

Different species are sometimes purposefully kept together in culture. Although half a dozen different introduced species are sometimes found together under the same log, possibly together with a native species or two, there are also centipedes, spiders, and other predators that it is likely one would not want to house with isopods. It is possible to keep different species of isopods together for relatively long periods, but if one species become overly numerous, it will eventually wipe out the other.

6
UNNATURAL SELECTION AND CAPTIVE STOCKS

BEYOND *ASELLUS* SP. purchased from Carolina Biological in the early 1990s and a 1995 group of *A. vulgare* that were supposed to be pill millipedes, my first foray into isopod culture began with an opportunity to acquire iridovirus-infected isopods from a friend who had acquired groups collected by a researcher in Spain in 1997. These sounded exciting as they were supposed to be beautiful. I had seen many "blue" invertebrates, including the small, gray Spanish millipedes. When these arrived I was floored. They even made electric blue crayfish look dull. The originals included a dozen blue *Armadillidium*, but they died after a few months and did not reproduce. The remaining animals were a *Porcellio* species.

There were more than a dozen blue *Porcellio* sp., four dozen solid gray, and a single orange animal. It took a few years and some selection before there were enough orange specimens to isolate a pure colony, since no offspring appeared orange until the second generation. I had attempted to isolate blue specimens a number of times but they always died after a few months. A few rare females would produce offspring before dying, but the resulting young were never born blue. Blue specimens would appear following a molt but did not change over until reaching maturity. Some would change over at initial maturity when they are not very large, though most change during late maturity at full size. Unfortunately, in isopods the iridovirus that causes the blue color appears to always be lethal to the individual after a few months, whether the animal is separated from the main culture or not. The culture was split into a few different cage styles and the blue remained a common sight (1-5% of adults) in the very damp cages, but grew less and less common in enclosures that were kept less consistently damp.

Purple *Porcellio* sp., the only such iridovirus infected orange form ever seen.

The main culture (versus the orange isolation) threw blue specimens with decreasing frequency from 1997 to 2008 when the very last one was seen. In 2006 a single orange specimen from the main, mixed culture changed to a handsome purple. I had previously believed the orange were immune since none had changed over in the last nine years and never within the pure-breeding orange culture. The mixed culture threw the rare variegated specimen, which was encouraged, but was far more rare than the orange. (In the middle years 5-10% would be orange, 90+% gray, and less than 1% variegated.) Through removal of countless gray and most orange, a nearly pure-breeding, variegated ("calico") stock was finally realized in late 2011 and perfected in 2012.

The iridovirus can infect *Armadillidium*, *Porcellio*, and probably most genera of terrestrial isopod. It is not easy to maintain in culture because it causes the death of the host, unlike the iridovirus that infects crayfish (McMonigle 2004). Also, only a small number of adults are normally infected. The high proportion of blue animals in my original group was because they were collected from a larger wild population. The *Porcellio* would have been lost quickly, like the *Armadillidium*, had only infected adults likewise been acquired. No visibly infected specimen ever lived more than a few months. Keep in mind that infected animals will be differently colored or patterned according to the individual and species, but animals with a tinge of blue or purple are not infected by this virus. A slightly purple *Armadillidium nasatum* cultivar (McMonigle 2013) and the naturally colored Costa Rican isopods can display a slight purple tinge, but they do not die and it has nothing to do with a virus.

At this point in time the major color morphs of any significance to the hobby all appear to be simple recessives. These are the orange forms of *Porcellio sp.* and *A. nasatum*, and the pied forms of *Trachelipus rathkii* and *Porcellio scaber*. The orange and pied form of *Armadillidium vulgare* and the white form of *Porcellio dilatatus* are probably similar, but they have not been isolated in culture. As simple recessives, this means if the male and virgin female are both orange, pied, or white, 100% of offspring will follow the parents. A single stock can take years to isolate when, as in most cases, there are only one or a few founding adults. This is because the life cycle is rather long and the first offspring will be heterozygous for the recessive. That is, they will all be the same phenotype (gray). If multiple unmated adults of the same recessive form are available, the isolation time can be reduced to nothing, though it may still be a few years for decent production. I have found it is not a good idea to try to isolate as quickly as possible because the genetic bottleneck can lead to weakly performing stock. Also, the recessive gene can be accidentally weeded out of the parent group and result in complete loss if the small, isolated group fails or fails to produce both genders. If the grays are removed slowly as the recessive becomes more common, many of the resulting offspring are more and more likely to be heterozygous over time. When there are many or mostly hets, a number of recessives will be available rather than starting with one or two founding animals. Of course the goal is to eventually remove all dominants and hets, but doing this too early on can actually cause it to take a longer rather than a shorter time to produce a viable stock.

Another possible recessive is a purplish form of *Armadillidium nasatum*. This form took only three or four years to isolate, while some variability between individuals was maintained. The hope was that once isolated the more deeply colored individuals would be selected until a truly purple animal could be isolated (a healthy natural form not prone to imminent death unlike the iridovirus infected specimens). Unfortunately after a few subsequent generations the intensity did not increase for any individuals. Even the best representatives of this form have proven rather unimpressive, except up close and in bright light.

The calico and halfwhite combinations are far more difficult to pin down genetically. The calico seems to be a combination of a recessive orange and another more variable recessive gene, because they show up only very rarely in gray and orange crosses. For years the culture would only produce a few of these compared to hundreds of oranges and thousands of grays. However, as the selected generations have progressed, the presence of resulting full orange offspring have grown more and more scarce and may have been eradicated. Removal of the dominant gray color genes and their disappearance from the stock over time makes perfect sense, but if all the adults were really co-recessive, around a quarter of the offspring ought to be fully orange. This seemed true early on when all three forms were present, but after many generations full orange were not even 1%. Additionally, the calico color form includes a number of different patterns and selecting for a variety of expressions is easily done, but difficult to explain genetically. I had originally been selecting for the gray splotches on an orange background, but the orange was often replaced with cream, tan, yellow, or a rare greenish so I switched to selecting for greater contrast rather than specific color. Bhella et al. (2006) believed variegation in *Porcellio scaber* (stock collected in Canada) was a sex-limited trait transferred by males and observable only on females, but with the solid-gray form entirely eliminated from the "calico" *Porcellio* sp. stock, it seems unlikely since it is probably the same gene (just as orange for various species is recessive). Their study into the genetics of color lasted only a few years and involved no completely isolated lines, so some of the conclusions were probable rather than concrete.

When it comes down to the *Porcellio laevis* halfwhite form, it would be far more difficult to plot phenotypes for the genetics involved because the same specimen can register as gray or white. *Porcellio laevis* produce a strange form called halfwhites, not because they are half white (though sometimes they are) but because a single specimen can be white one instar and gray the next. Grays can likewise change to white following a molt. It seems to be a recessive defect since it took a few generations to surface and after a number of selective generations nearly all specimens in the stock display this coloration, but each specimen is variable. In earlier generations, some of the affected animals would actually be partly or half gray and half white, but the majority would be white one molt and gray the next. Presently, specimens develop a gray blush after unspecified molts, and most specimens now are yellowish rather than white. The halfwhite could possibly be the same gene that causes white individuals in other species, since Bhella et al. (2006) believe albinism in *Porcellio scaber* was not due to inability to produce pigment but inability to transport pigment precursors into some tissues. Though displayed differently according to species, it is likely most recessive genes for coloration such as those for orange, white, or pied are the same in different isopods (and possibly even crayfish). The halfwhite may simply be the way in which the "pied" or "white" genes affect the color of *P. laevis*.

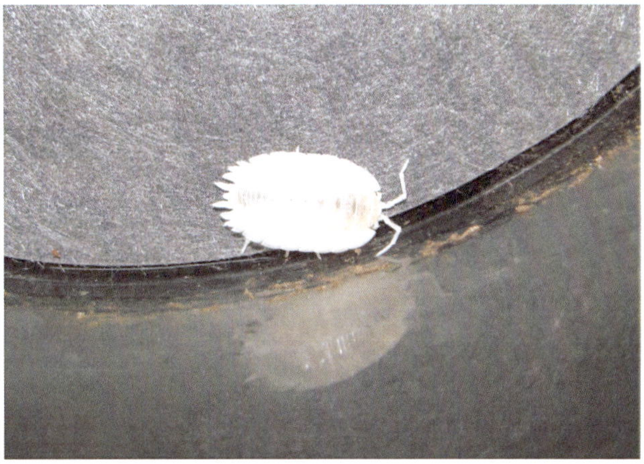

This white *P. dilatatus* originated from the "giant canyon" stock but is the only known white specimen produced so far. © Ron Wagler.

I tried to isolate a pale section *Porcellio* sp. (these might have been *Porcellio scaber,* but they were not as warty and grew as big as

Oniscus, but were not *P. dilatatus*) for at least a decade. Unfortunately even after all reproductives were isolated for this trait, only about one in a hundred offspring would follow. This certainly was not a recessive and so could not be isolated nor related to the gene causing the halfwhite form of *P. laevis*. After so many years producing a colony of maybe thirty, I grew frustrated and gave them away. I feel some regret since that keeper eventually killed them, but another two decades would probably make no difference as the prevalence never increased after isolating the reproductives.

Albinism is a genetic flaw in natural color schemes caused by an abnormal failure to properly synthesize dark pigment. White micropods and other naturally pale species are not considered even partial albino because white is the natural state. The piebald *A. vulgare*, *Trachelipus*, and *Porcellio* dalmatians, and pale section *Porcellio* sp. are all piebald forms, which are sometimes considered partial or incomplete albinos. Even among the whites I have not yet seen what I would consider a true or complete albino, because the few white forms I have seen have all had dark colored eyes. Albinos can be expected to have a dark line for the heart, but the eyes should not be dark. Even the superbly bright white form of *Porcellio dilatatus* has black eyes. Nevertheless, Bhella et al. (2006) report full albinos with white eyes do occur in *P. scaber*. It would be interesting to discover how the genes responsible for abnormal albinism in various species or pied forms are related to genes in more distant relatives like the abnormally white aquatic *Asellus* sp., other isopod species, and populations that are naturally white.

Morphological variations so far have played very little role in captive selection. The biggest reason is the primary character of interest, size variation, is almost impossible to isolate or select for. There is no ultimate molt and ultimate size is determined by external factors. One cannot simply isolate and breed the largest individuals within a reasonable amount of time. It takes years for sexually reproductive adults to reach full size and the eventual size is greatly affected by food and caging. Relatively tiny offspring will be producing young in the meantime, so specimens would need to be isolated and labeled. Such an endeavor would require decades of close attention to detail just to see if it might be possible. The "spiny" form of *Porcellio laevis* is the only morphological variation I am aware of that has been selected for. The outer edges of each body segment pull up and back at the sides rather than merging into a rounded outline.

One problem with creating a bunch of hets is, if culls are traded away as feeders or cleanup crews, it will give the competition a leg up. Since most of the work is in the initial buildup of hets, it is possible someone could take the culls and establish a perfected line more rapidly than the person who put in the first few years. Of course the loss is only bragging rights, since isopods are not worth much and looking

Vernier calipers are important for measurement since small sizes are difficult to estimate with any accuracy.

This purplish *A. nasatum* (left) is a natural form unrelated to the iridovirus but unfortunately it does not look like much from a distance.

"Calico" *Porcellio* sp. are the same bloodline as the orange and iridovirus specimens depicted elsewhere in the text.

into the possibility of variety patents is not even a pipe dream.

The most exciting form coming down the pipe from my perspective would be white *P. dilatatus*. It is a very easy to keep species that grows to double the mass of the orange *Porcellio*, but the only specimen seen so far from the "giant canyon" line died after producing only two het offspring. The *P. laevis* half-white and ruffled forms are interesting and have the potential to result in impressive, large cultivars, but it is a more sensitive species and these forms so far have been more fascinating than spectacular in appearance. Other interesting forms would be the orange and piebald forms of *Armadillidium vulgare*, but it is one of the slowest growing species so it might take a decade for isolation and good production. I have never seen the orange in person and the one piebald I managed to acquire was dead on arrival. Piebald might be especially lengthy to isolate for this species because it would be difficult to separate large immatures due to the variably patterned juvenile coloration.

MEASUREMENT

Measurements given are the maximum size for a known mature specimen, but breeding adults, especially for the largest species, can be reproductive at a third of this. For example, *P. dilatatus* can breed at just 6-7 mm (of 21 mm maximum) while *T. tomentosa* are reproductive around 2.5 mm (of 4 mm maximum). Given measurements were made using vernier calipers from the front of the head to the end of the telson. Uropods and antennae should never be included.

Due to the small size of isopods and the propensity for humans to make really lousy estimates of small invertebrates, a set of vernier calipers is suggested for those interested in discovering or reporting the size of isopods.

These three cultivars are similar in size. All three are large, older adults representing the usual maximum size.

If you measure your biggest "monster" specimen with vernier calipers you will be surprised when you find it does not measure as big as the maximum listed here, and these are conservative compared to most size claims.

ADVENTIVES

A number of the cultivars and species in the following lists are extremely common adventive animals introduced by Europeans hundreds of years ago to nearly every land area on earth. North American native terrestrial fauna is less than 2% of worldwide terrestrial species and primarily restricted to littoral and cave habitat. Most Americans have run across many isopods, but those who do not live near the ocean have probably never seen a native species. Natives listed here are *V. arizonicus*, a semi-aquatic sea slater, and the freshwater *Asellus*.

"Spiny" *P. laevis.*

Pale section form of an unidentified *Porcellio* sp. from Ohio.

Spanish infected blue and natural orange from from 1990s stock *Porcellio* sp., 2004. © Alex Yelich

Some of these orange *Porcellio* sp. do not diplay the dark line.

Dalmatian, *Trachelipus rathkii.* © Henry Kohler

Calico *Porcellio* sp.

Some listed species are neither naturalized nor native, like the *Porcellio* sp. collected in Spain and the Costa Rican jungle micropods. The origin of the striped dwarf stock is entirely unclear, while the Florida Keys *Venezillo* sp. might be native or adventive.

CULTIVARS

Porcellio sp. "Orange"

If you go back ten years or so, these were only traded as "Spanish Orange isopods" because the original bloodline came from stock originally collected in Spain. Sometimes the term neon or bright is thrown in prior to orange, since it is a very strikingly pigmented orange form (while orange forms of *Philoscia muscorum* and immature *Porcellionides pruinosus* are rather dull, etc.). These are a single-source bloodline so if you have some you can be certain they are descendents of the original female no matter what name they are sold under. It is a rather hardy stock that does very well, and older adults grow to 15 mm in body length. This is probably the most popular and useful of clean-up crew species. I have been trying to isolate the more solid coloration in which the orange color obscures the dark line of the dorsal abdominal artery, but after many generations and countless culls, less than a third look this way.

A number of people have labeled this stock *Porcellio scaber* not because of a detailed genitalia assessment nor identification by a specialist, but rather because they assume it was collected in North America where it would likely be the only real option. However, since this bloodline was collected in Spain, it could easily be a different species or subspecies. A group of scientists who studied the genetics of color forms using introduced North American specimens of *P. scaber,* found their orange specimens "were always seen with a white edge" (Bhella et al. 2006). Their *P. scaber* looked rather different and over the course of the study they were unable to isolate a pure breeding line.

Porcellio sp. "Calico"

This stock is a single-source bloodline, but the quality of the strain will depend on how recently it was separated from the parent stock since previous generations had the ability to revert somewhat. Recent generations are not only fully, or nearly true breeding, but have become better and better looking (due to selection for color). Although a specimen's basic color does not change, full coloration is often not visible until the animal is mature or has reached full size. One unusual variation is transparent in parts so it is possible to see halfway through sections of the animal. This is the exact same bloodline as the *Porcellio* sp. Orange, but split from the parent stock in the late 1990s. (After all these years the strain is still in process, since "calico" is not a simple recessive.) This strain seems a little hardier than its counterpart and also can grow as large as 15 mm.

Peach pillbugs.

Some of the peach pillbugs are almost white in color.

"Original Dalmatian" *Trachelipus rathkii*
This is the Original Dalmatian form that has been available in very limited numbers since 2010 (first labeled as *Porcellio* sp.). It is a beautiful white creature with black eyes and variable black spotting. These look very much like *Porcellio scaber* but there are five white patches (lungs) along each side of the abdominal pleopods. Also, it behaves differently. It is prone to running quickly, jumping, and tucking in all the legs when disturbed, so it is a bad idea to hold cage decorations above anything but the cage. I collected one specimen in northern Ohio in 2008 and it was a few years before there were a dozen. It is a very beautiful form, but the stock has proven somewhat less productive than the standard gray form, and the number of specimens after more than five years can probably still be measured in the hundreds, partly due to trading off too many early on.

The *Porcellio scaber* Dalmatian is more variable than the *T. rathkii* in type, color, and intensity of spots (spots are variable in placement for both species).

"New Dalmatian" *Porcellio scaber*
These can look similar to the Original Dalmatians, but usually have larger splotches that look like coffee stains and extra colors or very little color. (The name Calico is already taken by something that looks a lot different.) Specimens do have two large lungs on each side of the pleopods like all *Porcellio*. Although the coloration is less striking, it seems to grow and breed more quickly than the *Trachelipus* Dalmatian. The first generation adults were collected in 2012 from among a normal wild population in Michigan. They immediately bred true when isolated, which would have been rather amazing if the females were fully mature, since isopods can retain sperm for long periods.

"Peach Pillbug" *Armadillidium nasatum*
This "orange" form is more of a peach color and retains the light and dark banding of the normal gray form. I originally found one adult in the woods behind my house in northern Ohio. I also collected one in south central Pennsylvania and another in eastern Virginia. Despite a good number of founding adults (three), I have been working on getting and crossing hets since 2009 and only recently started producing a decent number of peach early instars. The intensity of the orange is variable, but none have included gray coloration. One large immature specimen is so pale it is nearly white and a few smaller specimens also appear abnormally white. Adults will breed year round, but immatures can take a year or more to reach maturity, so it will take years to see how the white forms develop.

"Halfwhites Giant Canyon" *Porcellio laevis*
Halfwhite stock has been available for a few years, but previous offerings were not completely isolated and this species' variability by

The original Dalmatian, *Trachelipus rathkii.*

As of 2013 the halfwhites are primarily isolated but number less than a hundred. Coloration and intensity seem less variable but there is a bit of brown to yellow coloration.

This photo from 2011 represents three or four years of selecting for the halfwhite form of *P. laevis*.

development is difficult to follow. Isolated, selected stocks presently throw mostly or entirely white, but the white has a gray blush or turns yellow variably. Specimens of this stock hail from a canyon in southern California and so are the large form of this species. "Spiny" specimens have appeared in some cultures.

STANDARDS

Armadillidium maculatum

Zebra Pillbug

Clean-up: This seems to be a reasonably easy to breed and hardy pillbug but it has not yet been tested in terraria. Size: Available stock grows to a similar size of the following two species, around 14mm, maximum size uncertain.

This rather spectacular pillbug is so new it was included in the small guide version only as a future possibility or dream species. Most of the specimens are dark with light margins similar to *G. marginata* pill millipedes though the bands are broken up variably in different specimens. It seems to be hardy, has already begun to reproduce in some numbers, and will probably be established quickly in culture.

Armadillidium nasatum

Large Nose Pillbug

Clean-up: This pillbug is hardy and suited to a far greater variety of terraria than *A. vulgare* because low ventilation and high moisture are usually accepted, though they do just as well in drier habitats. Size: This species is often listed as the largest species in the genus (maximum 21 mm versus 18 mm for *A. vulgare*) but finding an old adult larger than 15 mm for either species is exceptional to the point of obscurity.

Small specimens often stay in a ball, but adults rarely can be coerced to stay closed up. This species is fully capable of drawing in and completely sealing antennae within the sphere like the more familiar *A. vulgare*. Females can produce incredible numbers of mancae, but these often stay rather small for months and months. Specimens are normally striped, but solid, dark gray animals are not difficult to find. Members of this genus are often associated with disturbed areas like roadsides and railway cuttings and show a marked preference for areas with limestone in nature (Nichols et al., 1971). Captive culture requires no limestone and presence of limestone in the enclosure shows no measurable effect on productivity or growth.

Armadillidium vulgare

Common Pillbug

Clean-up: This species works for clean-up in well-vented terraria, but it is not very productive nor an aggressive feeder. Size: 17 mm (Spanish bloodline maximum); few northeastern U.S. specimens measure greater than 14 mm.

A friend of mine who owned a pet shop said he had a diplopodan gift for me in 1995. I

Armadillidium maculatum.

This species is normally slate gray and checkered with lighter markings, *A. nasatum.*

Cylisticus convexus. © Henry Kohler

stopped in to find a clear 8″ tub containing some recently imported "blue millipedes" and supposed "pill millipedes" from Spain (country of origin being sheer coincidence, as there is no connection to the Spanish orange *Porcellio* or *P.* cf. *bolivari*). The "blue" creatures were actually gray and somewhat small *Ommatoiulus rutilans*, while the pill creatures were only *Armadillidium vulgare*. Although they were some of the largest *A. vulgare* I had ever seen, I was disappointed by the small millipedes and rare pillipedes that turned out to be common pillbugs. Not one to look a gift horse

Adult *A. vulgare* are usually a slate gray with some yellowish markings.

in the mouth I took them home. They stayed in the small tub for a few years, were moved to a ten-gallon for a number of years and were moved to a plastic shoebox five or six years ago. Large numbers died after the last move due to restricted ventilation. Most species do not require ventilation beyond the gap around the lid. The problem was recognized before it was too late and some holes were drilled in the lid. The colony eventually rebounded. A few years back the shoebox was stacked under an *A. nasatum* culture and one day I noticed there were a lot of "extra" *A. vulgare*. I was worried they would suffer losses since the *A. nasatum* are faster growing and far more virile. Small to medium immature *A. vulgare* are banded and variable in color and so can be very difficult to differentiate by pattern. The young often stay in a ball even if I had the time and patience to check each tiny critter's face with a 10x loupe magnifier. A few days later I realized it was going to be easy to pull out the *A. nasatum* because I was looking at the wrong thing. Of the two, only the *A. vulgare* are glossy.

Adults are normally solid gray or gray with yellow markings. Immatures are highly variable and usually striped and spotted and can have large pale sections, but still end up looking similar at maturity. In Europe there are a number of smaller *Armadillidium* species, but most would be very difficult to differentiate from the variable *A. vulgare* immatures. A few of the more colorful species like *A. klugii* (from the eastern shores of the Adriatic and Ionian Seas) are thought to mimic the bright color patterns of certain widow spiders, *Latrodectus*, but no research has been done on this longstanding theory (Schmalfuss 2013).

Overall, *A. vulgare* is an easily cultured species as long as ventilation is not restricted and time is not of the essence. When adults finally breed, they do so in volume (up to 100 per female), but then the young take up to two years to mature.

Asellus sp.

AQUATIC ISOPODS

Clean-up: Extremely limited even in aquatic setups. Size: 15 mm.

The four antennae are obvious and the body looks nothing like other familiar isopods. The ones purchased from Carolina Biological twenty years ago lived for a while, but I did not feed them dead leaves and recall no offspring. A few years back I picked some that were collected in Michigan. I have kept them going for a few years in a 10-gallon tank filled with aged tap water 3/4 to 4/5th to the top. No filtration is used. I have added them to a few other cages with sponge filters and each time they died after a few weeks or so. Food consists of rotten, brown hardwood leaves and sinking fish pellets or flakes. Fish food is fed sparingly as without filtration it would be very easy to poison and kill off the entire culture. Leaves take a few days to a few weeks to sink to the bottom where they can be fed on. Captive care is low maintenance, but it is important to make sure fish food that is not eaten in two days is siphoned out. It is amazing both how little a few thousand animals can eat and how difficult it can be to see them among a handful of leaf litter. They take a year to reach maturity and then may not live even six months afterwards. If no mistakes are made in husbandry there can still be an unsettling amount of die off of the larger animals (probably adults dying of old age, but adult size is highly variable). They are not very interested in eating the dead bodies of cage mates, unlike some of the terrestrials. For this reason, dead animals can be obvious and numerous at certain times of year. It is important to siphon out the frass every few months because a surprisingly thin layer can lead to anaerobic growth, associated smell, and tons of dead animals, including immatures. Another problem is difficulty in rearing monster males. The wild males were spectacular, but I have not seen anything nearly so big yet. There is a recessive white form of the species pictured, but I have not yet been able to acquire stock.

Asellus sp. original stock from Michigan.

These two large *C. convexus* had been breeding females when smaller but seem to have male features such as elongate uropods and oversized basal pleopods.

Cylisticus convexus
SKINNY PILLBUG
Clean-up: Useful in a variety of terrarium types. Size: 14 mm.

This pillbug species is far less widespread than our adventive *Armadillidium* species. It is much faster moving and at no size does it ever seem to stay in a ball. It readily rolls into a ball, but does not retract the antennae and has long, spiked uropods. This species is very hardy and constantly reproduces, but mancae are tiny and can take more than eight months to reach maturity.

Desert form stock originally collected in Arizona with *Arenivaga* sp. desert cockroaches, *P. laevis*, (Photo taken 2004). © Alex Yelich

Ligia exotica manca.

Ligia exotica possess exaggerated biramous uropods and highly articulated, long antennae.

Ligia exotica

Wharf Slater

Clean-up: Possibly very useful in certain brackish paludaria. Size: The largest specimen of fifty collected in southern Florida was 26 mm. Larger measurements seen listed may include appendages.

Also known as the wharf roach, this introduced species is found across warmer areas of the U.S. east coast, as well as Europe and Asia. The wharf slater is an amazingly fast and spectacularly aware creature. Six were initially placed in my reef tank to see how they would react to a purely aquatic marine environment. They seemed at home and remained underwater for six hours before I found the first one up on the side. Unfortunately the Bangai Cardinal fish and some hermit crabs were seen chasing them around, so the *Ligia* were removed as they could be found over the next few days. One had lost both long, branched uropods in the tank (probably eaten)—it molted just two weeks later and regenerated them almost perfectly.

Specimens have a hearty appetite and will eat their own mass in freshwater algae in 24 hours. Marine hair algae and *Caulerpa* spp. are also eaten, but these algae must sit out of water for a few days till dead. They will feed on nearly anything other isopods will eat and are especially fond of fish flakes. Most pictured specimens have been kept on an uneven layer of play sand drenched in artificial seawater and seem easy to keep short term. They have molted countless times and produced mancae. Unlike most terrestrials, the exuvium is not eaten immediately, but often left for a few hours. Like other *Ligia* species the males are the far larger gender and have enlarged tibias on the front three leg pairs. It is easy to sex specimens from above by looking at the legs, even in animals as small as 8 mm. Eggs develop rapidly and 1 mm long mancae are released three weeks after eggs are observed within the marsupium. Mancae do not start out pale or white, but are similar in color to adults. It is a good idea to add extra shelters or remove mancae, as many can be trampled by the adults.

Ligia pallasii

Pallasi's Giant Northwestern Slater

Clean-up: Difficult to keep alive, no value as a cleaner. Size: 35 mm.

At this size, *L. pallasii* could be the only terrestrial large enough to make a good pet, but it may be too difficult to culture and few have tried. This is a rather nice-looking species, as the immatures are a handsome variegated green and white pattern, while adults transform into a brown version of army green. The first few times I tried to acquire specimens they were all DOA, but I was able to keep a group of eleven growing, molting, and somewhat healthy for half a year in 2008. However, between months four and six, one specimen would die about once a week. They were kept in a 5-gallon bucket with a screen lid on a sand substrate kept to the level with artificial saltwater so that a few areas of the sand could be dug out and offer dipping areas for the gills.

Ligia pallasii adults can be rather large.

Oniscus asellus common adult coloration.

Evaporation was replaced with freshwater, or the bottom would have become a solid block of salt. I began offering various marine algae including *Caulerpa* spp., *Valonia* spp., two different red macroalgae, slime, hair algae, and prophase kelp, but even the last was met with almost zero interest. Various fruits also appeared untouched. The only things they fed on with much interest were dead hardwood leaves, but Java moss, dog food, and fish flakes showed limited signs of feeding. My goal was to see how long they could be kept alive, not how quickly they die at varying temperatures, so they were kept between 69°-74° F. Considering their northwestern location, anything above 77° F for long might be deadly.

Oniscus asellus

Clean-up: This species may be useful as a clean-up crew in cases where reproduction is not desired, but it can die from warm temperatures. Size: 17 mm; near maximum specimens are not difficult to find.

This very common species is usually the largest species seen in most areas. It looks similar to *Porcellio* species but the outline is less streamlined, the flagellum is composed of three articles, and there are no visible lungs on the pleopods. This is one of the slowest growing species with specimens reaching maturity only after about two years (Nichols et al. 1971). It is easy to keep adults alive for a few years, but those who report seeing babies produced in captivity almost certainly collected a few *Trachelipus* or *Porcellio* with their *O. asellus*.

Philoscia muscorum

Fast Isopod

Clean-up: It does not make a good terrarium clean-up species since specimens tend to die when hot and wet, but they also do not like dryness. Size: Aged specimens seldom exceed 10 mm in length.

Philoscia muscorum brown and orange-red forms do not look terribly different.

The name refers to its rather impressive land speed since it tends to run rather than stay put on discovery and is capable of running more quickly than the average isopod. This is another commonly encountered, introduced species from Europe. Adults can be somewhat colorful as individuals may display red, yellow, and orange in addition to brown and gray. The body color is often offset by a contrasting margin of red or white. A wide array of colors are common in groups of wild-caught adults, but the colors are not nearly so bright nor as pretty as they are for cultivars of other species. It is

This is usually as big as a terrestrial isopod can get, a little bigger than a pumpkin seen, *P. dilatatus*.

Porcellio dilatatus. © Henry Kohler

easy to keep specimens alive and healthy in captivity for two years or more, but captive production of immatures is more theory than fact.

Porcellio dilatatus "Giant Canyon"
Clean-up: Specimens work well for clean up in a variety of cage types, but can be rather conspicuous due to large size. Size: Old specimens are usually from 17-18 mm, but I just measured one at 21mm. Stock collected elsewhere may not grow so large.

Although this is not a rare species, the stock traded in the hobby is probably all from a single source, a canyon in southern California. It is a large, hardy, reproductive stock that is resistant to dryness and survives high humidity. It takes three or four years for the largest specimens to develop in a colony, but these are very impressive, bulky creatures. Medium to large specimens are usually a monochromatic gray, sometimes with red-brown edges, but large, old specimens can have pale splotches. It is a burrowing species, with specimens often found an inch deep in fine substrate. The curl defense is highly pronounced and sometimes specimens will stay this way for minutes.

Porcellio laevis
SMOOTH PORCELLIO
Clean-up: These are not terribly voracious, but accept varying conditions and can be easily monitored since they are motile and rarely burrow. Size: The largest adults in an old culture often reach 20 mm.

I have maintained two forms of this species from North America that have proven vastly different in behavior and development. The first was acquired from mesquite scrub in the dry hot Sonoran Desert in Arizona circa 1998 and kept though 2006. Specimens were more resistant to desiccation than almost any other animal I have kept. They burrowed readily in substrate and would retract the legs and remain motionless. The largest specimens seldom reached 16 mm after many years, and there was no difference seen in the male and female uropods. The second is the base stock for the Giant Canyon Halfwhites and came from the same canyon in southern California as the "giant canyon" *P. dilatatus*. This habitat seems to offer selective pressure for large size, since wild-caught specimens are often impressively large. Adults are the largest fully terrestrial isopod in North America and 20 mm in length is commonly achieved by old adults. Old males have long uropods that are double the length of similarly-sized females. This form rarely hides in the substrate with the legs retracted.

This 20mm *P. laevis* from southern California does not look terribly large next to an oversized Julid and massive Spirobolid millipede from the same locale.

Porcellio scaber
ROUGH ISOPOD
Clean-up: Useful in various settings. Not nearly as accepting of warm temperatures as the cultured *Porcellio* sp. orange and calico stocks. Size: 14 mm.

This is supposed to be one of the most common adventives. It is easy to locate in many areas of the United States, but in my experience

These *P. pruinosus* possess exoskeletal surface structure that gives them a chalky appearance from most angles.

Normal gray coloration of *Porcellio scaber.* (CC-SA) D. Sikes

Whether dark or light in color the head of this species tends to be darker than the body, *P. spinicornis*.

Common coloration, *Trachelipus rathkii*.

it is found in far, far lower numbers than *Armadillidium, Oniscus, Philoscia, Trachelipus*, etc. It is hardy and easy to keep, but not as virile as some commonly kept species.

Porcellio spinicornis
SPINE HORN ISOPOD
Clean-up: It is seldom hardy enough to be useful in terrarium clean-up. Size: 15 mm.

The name comes from the small spine found on each antenna. This species produces readily and is similar to other *Porcellio* in care. It can be confused with other *Porcellio*, but the head is usually darker than the body and there is a nearly microscopic spine at the end of the third antennal segment. Some forms are colored just like the usual *Oniscus asellus* with a white border and irregular greenish yellow markings down the back (that are also seen on many adult *A. vulgare*). *Porcellio scaber* forms that are gray with white edges usually do not have the greenish markings. Stocks can be solid gray or speckled like the ones shown here. This species usually avoids high humidity and is often found under rocks in drier areas.

Porcellionides pruinosus
CHALKY ISOPOD
Clean-up: These can work well in various cage types and do not normally burrow. Size: Large, old adults seldom reach 9 mm.

The most interesting aspect is the black-and-white banding of the antennae, though sometimes the uropods are orange as well. The body of the adult is gray but has a milky or chalky look from different angles. Immatures have the same surface texture, but are a pale, pinkish orange up to a relatively large size. The exoskeleton is pliable and does not crunch audibly like most species when accidentally damaged. Specimens are rather quick growing and can reach maturity in two months. The number of young per brood seems low but they reproduce more rapidly than most other species and are simple to care for.

Trachelipus rathkii
WETLAND ISOPOD
Clean-up: Useful species that survives a variety of caging extremes, even somewhat dry terraria. Size: 13 mm.

These look a lot like *P. scaber* and likewise come in a plethora of different color forms, but a familiar person can often tell the two apart by pattern, since each of the different forms tend to be slightly different in pattern. The bumps on the body surface tend to be less pronounced, but this is easier said than seen. The pleopod lungs, five down each side like *Cylisticus* and *Venezillo*, are the only easily seen feature to differentiate it from *P. scaber*. It is a more skittish species which is prone to short bursts of speed when disturbed. Specimens are as hardy as the sturdiest *Porcellio* species and reproduce well under varying conditions, but are not so voracious and usually do not eat their own dead. This species is a widespread adventive, but most often found in areas that are flooded at regular intervals, such as close to rivers and creeks, freshwater marinas, and wetland or bog areas.

NEW SPECIES

There are a few fantastic looking species that have been rumored in the hobby, though whether these can be reared with any success and exactly how big they are or how big they would get in captivity are uncertain. Accounts of captive reproduction are sketchy at best. Maximum size in nature may be difficult to match in captivity. Also, these may not have been and may never be in the United States. These include *Helleria brevicornis* from Corsica, a pillbug that can grow to twice the mass of the largest *Armadillidium* sp., and two very handsome white and yellow species collected in

Captive-reared first-generation female *Porcellio* cf. *bolivari*. © Ingo Fritzsche

Wild-caught male *Porcellio* cf. *bolivari*. © Ingo Fritzsche

Nagurus sp. striped dwarf.

Spain, *Porcellio expansus*, and *Porcellio* cf. *bolivari*. This last one is a handsome creature with lemon yellow painted down the back and very long male uropods. It is presently being kept with limited success by one person in Germany. Thirty were collected from Spain, Costa Blanca near Benidorm, 2012, and the group grew to about 120, but they have proven difficult and die easily from stress. About 100 died, many when fed a grass they did not like, but twenty are doing well. One adult male is alive after a year and young have taken a year to mature (pers. comm. Ingo Fritzsche).

DWARFS

Nagurus sp.

STRIPED DWARF

Clean-up: Excellent for most terraria, but climbing propensity should be considered. Size: 6.5 mm

This is a unique looking species with its longitudinal stripes and white telson. It really likes to climb and if the sides and top of a shoebox cage are a little dirty, large numbers can be found hanging under the lid. Care must be taken when popping the lid off. Specimens were able to survive the cold temperature 38° F (3° C) of a refrigerator for four days, but extended or deep cold should be avoided. This stock, mostly likely single-source, has been traded in the dart frog hobby since at least 2009.

The head shape is unique to the genus *Nagurus*. © Henry Kohler

Atlantoscia floridana

FAST DWARF

Clean-up: Good for damp terraria. Size: Up to 8 mm, but in a culture only a small percent ever seem to grow so large.

These look very similar to the common adventive *Philoscia muscorum*, but large, old adults are only half the mass of a small, young adult *P. muscorum*. The two most significant differences are temperature and reproduction. They die if placed in a refrigerator and require tropical temperatures for survival, while *P. muscorum* survive well below freezing.

Striped dwarf on a potato chunk.

Full-grown micropill *Venezillo parvus* (top) compared to mature Dwarf Key *Venezillo* sp. Standard pillbugs are even larger by comparison.

Fast dwarf.

Venezillo sp. from the Florida Keys.

Venezillo arizonicus © Peter Clausen

Females are highly productive and reproduce year round in captive enclosures. The telson is shaped differently, there are three microscopic setae at the end of each antenna, and the body is more narrow, but the pattern of spotting is similar. However, unlike *P. muscorum*, this species so far has shown very little variation in the adult coloration. From common viewing distance on a dark background they look silver, but on close inspection the color is chestnut with cream markings. Present stock is from a single source and may date back as far as 2002. Like its larger cousin, this species scatters and runs rapidly when exposed, but it is much faster.

Venezillo sp.
Key's Dwarf Pillbug
Clean-up: Probably useful, but with slow reproduction and limited voraciousness. Size: These dwarf pillbugs rarely exceed 8 mm.

There are at least seven small native members of this genus and one or more adventives. They are only found in warm, southern areas of the country, but habitats can range from the scorching desert for *V. arizonicus* to subtropical wet areas of islands off Florida. None have yet been available commercially, but this handsome species from the Florida Keys is established in culture. It may prove useful for terraria unsuitable for the larger pillbugs, but probably could never be a feeder since a female takes months to produce half a dozen mancae.

MICROPODS

Unknown sp.
Jungle Micropod
Clean-up: Excellent, as they often stay out of sight and reproduce well in most terraria. Size: Old adults rarely exceed 3 mm.

Older immatures and adults usually appear to be a sort of purplish gray, especially from a distance, giving rise to the name "purple

Jungle micropods range from slightly orange to slightly purplish gray.

Jungle micropod close up.

micros" on many sales pages. Some are a shade of orange which might be attributed to the orange eggs, but radiates throughout the whole body. This is the only species which almost never holds onto wood pieces, so specimens have to be transferred or removed from the substrate by hand. In many ways this species is similar to *T. tomentosa*, but not so much to assume it is in the same genus or even closely related. This stock has a single-source origin, as it arrived from Costa Rica in 2007.

White micropods.

Trichorhina tomentosa

White Micropod

Clean-up: Specimens often stay out of sight and reproduce well. These are not as productive as jungle micropods under certain conditions. Size: Old adult specimens rarely exceed 4 mm in body length.

This species is found in Central and South America, but was long ago introduced to Florida and is commonly found in greenhouses worldwide. If a culture source cannot be found, a trip to a few local greenhouses is surprisingly likely to yield results. I originally acquired specimens in 1999 and since that time they have become one of the most common and desired species for use in dart frog terraria. This species burrows (usually less than half an inch deep).

Although a female only has half a dozen young at a time, there are no specimens in a culture that cannot have young (as reproduction is through parthenogenesis).

Venezillo parvus.

Venezillo parvus

Maculated Micropill

Clean-up: Good for wet and humid tropical terraria. Size: These qualify as "micropills" since they make the other *Venezillo* species pictured here look massive. Maximum adult size is 5 mm.

This handsome, tiny species is adventive to subtropical and tropical areas around the world, including Florida, Georgia, and Hawaii. The specimens pictured are from southern Florida and come from a moderately large (a few hundred plus individuals), two-year-old, captive colony. About a quarter of the specimens are orange, but the orange form is a rather dull morph.

White micropod close up. © Henry Kohler

Jungle micropod close up. © Henry Kohler

Haplopthalmus sp., Medina County, Ohio. This temperate micropod is smaller and thinner than *T. tomentosa*.

Porcellionides pruinosus .

GLOSSARY

ADVENTIVE: Naturalized or introduces species.

BIRAMOUS: Dividing into two main trunks or branches.

BUNCHING: Clustering together during dry periods to reduce water loss.

CRYPSIS: Mimicry of common background objects like rocks or plants rather than fauna.

DESICCATE: To dry up or dry out; dehydrate to the point of death.

DIMORPHIC: Possessing two distinct forms or shapes. Sexually dimorphic males and females are visibly different in shape. This is often expressed in the pleopods and uropods of isopods.

EXOCUTICLE: Outer layer of the exoskeleton.

EXUVIUM: The shed exoskeleton.

FLAGELLUM: The terminal segment of the isopod antenna normally composed of two to twelve articles, the number of articles can be useful in identification. There are no muscles in this section of the antenna.

FRASS: The solid waste of invertebrates.

FURCULA: jumping mechanism of springtails.

HET: Short for heterozygous, refers to an animal with one dominant and one recessive gene for a specific trait. Hets look the same as an animal with two dominant genes. If crossed with another het the theoretical offspring will be 50% het, 25% dominant, and 25% recessive.

HYPOPUS: A unique stage of certain types of mites that is a response to reduced food, they molt into a suction cup stage and search out exoskeletons of living arthropods to cling to.

INSTAR: Nymphal stages after hatching and between molts (i.e. a third instar nymph would have molted twice).

IRIDOVIRUS: Here: Isopod iridescence virus, it results in blue coloration of a percentage of adult isopods but has always proven lethal within a few months.

Ligia exotica feeding on freshwater algae.

LITTORAL: Of or living within the area between high and low tide.

MACROALGAE: Algae that resemble vascular plants; seaweed.

MANCA (pl. MANCAE): 1st instar isopod. This stage is nearly always white in color.

A volvating zebra pillbug.

MARSUPIUM: Specialized egg pouch between the front legs and comprised of oostegites.

MAXILLIPED: Outer, third pair of mouthparts that arise from the thorax.

MEDIUM (pl. MEDIA): The food as well as substrate used in culturing fruit flies, bacteria, etc.

OCELLUS (pl. OCELLI): Light detection organs.

OMMATIDIUM (pl. OMMATIDIA): structural unit or complete facet of the compound eye.

OOSTEGITES: Plates that form the marsupium and arise from near the base of the pereopods.

PALUDARIUM (pl. PALUDARIA): A terrarium that includes both terrestrial and aquatic areas, usually half and half.

PARTHENOGENESIS: Development through unfertilized eggs.

PEREON: Middle section of the isopod body, thorax as separate from the head with incorporated maxilliped segment and the pleon.

PEREOPODS: Ventral uniramous appendages attached to the pereon. These are the walking legs for most species though they can be useless for locomotion and adapted for filter feeding or prey capture in some marine isopods.

Venezillo sp. from the Florida Keys.

Some populations of *A. vulgare* throw an orange form. Collected in Oregon. © Peter Clausen

PHENOTYPE: The appearance of a specific dominant trait such as gray body color. An animal that is heterozygous will appear the same as one with both dominant genes.

PLEON: Abdomen as separate from the pereon, contains the pleopods, telson, and uropods.

PLEOPODS: Biramous appendages on the pleon that form the gills and can be adapted in relation to mating for the males, known as swimmerets for many aqautic isopods.

PLEURITES: dorsal plates of the pleon not including the telson.

SETA (pl. SETAE): Thin bristle or hair-like extension of the arthropod exoskeleton.

TELSON: Terminal body segment or tail. This is part of the pleon and may be called a pleotelson to differentiate it from the telson of a scorpion.

UNIRAMOUS: Nonbranching.

UROPODS: Biramous or uniramous appendages on either site of the telson, stylets. These form the paddles of the tail in swimming isopods, lobsters, etc.

VENTRAL: Bottom or underside.

VOLVATING: The property of being able to curl into a spherical shape.

Fast dwarf isopod eating freshwater algae.

Rosy woodlouse, *Androniscus dentiger*, from the U.K. (CC-SA) Gilles San Martin

BIBLIOGRAPHY

Allsop, David J. (2003) *The Evolutionary Ecology of Sex Change*. The University of Edinburgh.

Bhella, S., et al. (2006) Genetics of pigmentation in *Porcellio scaber* Latreille, 1804 (Isopoda, Oniscidea). *Crustaceana* 79(8): 897-912

Caldwell, Robert. (2008) Giant deep-sea isopod (*Bathynomus giganteus*). *Invertebrates-Magazine*. 7(4): 19-20.

Chiao, C. C., T. W. Cronin, and N. J. Marshall. (2000) Eye design and color signaling in a stomatopod crustacean, *Gonodactylus smithii*. *Brain, Behaviour and Evolution* 56: 107-122.

Dunn, Gary A., ed. (1993) *Caring for Insect Livestock: An Insect Rearing Manual*. Special Publication No. 8, Young Entomologist's Society, Lansing, MI.

Forster, R. R., and L. M. Forster. (1980) *Small Land Animals of New Zealand*. John McIndoe Ltd., Dunedin, New Zealand.

Fritzsche, Ingo. (2013) Auf wirbellosenpirsch in der Levante—Spannende entdeckungen an der Sonnenkuske Spaniens. *Bugs Das Wirbellosenemagazin* 1(1): 32-36.

Hellweg, Michael R. (2009) *Raising Live Foods*. TFH Publications, Neptune City, NJ.

Jass, Joan, and Barbara Klausmeier. (2001) *Terrestrial Isopod (Crustacea: Isopoda) Atlas for Canada, Alaska, and the Contiguous United States*. Contributions in Biology and Geology. Milwaukee Public Museum.

Kight, Scott L., M. Martinez, and A. Merkulov. (2001) Body size and survivorship in overwintering populations of *Porcellio laevis* (Isopoda: Oniscidea). *Ent. News* 112(2): 112-118.

Linsenmair, Karl E. (1974) Some adaptations of the desert woodlouse *Hemilepistus reaumuri* (Isopoda, Oniscoidea) to desert environment. Verhandlungen der Gesellschaft für Ökologie 4: 183–185.

Linsenmair, Karl E. (2007) Sociobiology of terrestrial isopods, in: *Evolutionary Ecology of Social and Sexual Systems: Crustaceans as Model Organisms* (J. Emmett Duffy and Martin Thiel, eds.).

Kneidel, Sally. (1994) *Pet Bugs: A Kid's Guide to Catching and Keeping Touchable Insects*. John Wiley & Sons, Inc., New York.

Masters, Charles O. (1975) *Encyclopedia of Live Foods*. TFH. Publications, Neptune City, NJ.

McCormick, Stephanie. (2007) Arthropoda 2: Subclass Malacostraca. (Online paper for Marine Biology 377, Oakland University, Rochester, MI). http://marbio.homestead.com/Arthropoda2_07.doc

McMonigle, Orin. (2001) *Assassins, Waterscorpions & Other True Bugs: Care and Culture*. Elytra and Antenna, Brunswick, OH.

McMonigle, Orin. (2003) *Giant Centipedes: The Enthusiast's Handbook*. Elytra and Antenna, Brunswick, OH.

McMonigle, Orin. (2004) Terrestrial isopods. *Invertebrates-Magazine*. 4(4): 9-12.

McMonigle, Orin. (2011) *Invertebrates for Exhibition: Insects, Arachnids, and Other Invertebrates Suitable for Display in Classrooms, Museums, and Insect Zoos*. Coachwhip Publications, Landisville, PA.

McMonigle, Orin. (2013) The beauty of volvation and the pretty peach pillbugs. *Invertebrates-Magazine* 12(June):2.

Mocquard, J.P., A. Pavese, and P. Juchault. (1980) Seasonal reproduction determinism of *Armadillidium vulgare* females (Crustacea, Isopoda, Oniscoidea): 1. Action of temperature and photoperiod. *Ann. Sci. Nat. Zool. Biol. Anim.* 2(2): 91-98.

Nichols, D., J. Cooke, and D. Whitely. (1971) *The Oxford Book of Invertebrates*. Oxford University Press, Manchester.

Schmalfuss, Helmut. (2003) World catalog of terrestrial isopods (Isopoda: Oniscidea). *Stuttgarter Beiträge zur Naturkunde, Serie A (Biologie)* 654.

Schmalfuss, Helmut. (2013) Revision of the *Armadillidium klugii*-group (Isopoda: Oniscidea). *Stuttgarter Beiträge zur Naturkunde A, Neue Serie* 6: 1-12.

Souty-Grosset, C., et al. (1994) Interpopulation variability of the seasonal reproduction in the terrestrial isopod *Armadillidium vulgare* Latr. (Crustacea, Oniscidea): A review. *Act. Oeco.* 15: 79-91.

Tait, N., R. C. Vogt, and H. Dingle, cons. (2005) *The Encyclopedia of Reptiles, Amphibians & Invertebrates*. Weldon Owen Pty Ltd. Sydney, Australia.

Walls, Jerry G. (1982) *Encyclopedia of Marine Invertebrates*. TFH Publications, Neptune City, NJ.

Werner, F., and C. M. S. Olson. (1994) *Insects of the Southwest*. Fisher Books, Tucson, AZ.

Dalmatian *Trachelipus*.

Coachwhip Publications

CoachwhipBooks.com

INVERTEBRATES FOR EXHIBITION
Orin McMonigle
ISBN 1-61646-105-5

MILLIPEDS IN CAPTIVITY
Orin McMonigle
ISBN 1-61646-143-8

www.ingramcontent.com/pod-product-compliance
Ingram Content Group UK Ltd.
Pitfield, Milton Keynes, MK11 3LW, UK
UKRC040951141026
15625UKWH00083BA/673

* 9 7 8 1 6 1 6 4 6 2 0 7 9 *

9 781616 462079